Science and Technology in the Academic Enterprise: Status, Trends, and Issues

A Discussion Paper

The Government-University-Industry Research Roundtable

National Academy of Sciences
National Academy of Engineering
Institute of Medicine

2101 Constitution Avenue, NW
Washington, DC 20418

National Academy Press Washington, D.C. October 1989

THE GOVERNMENT-UNIVERSITY-INDUSTRY RESEARCH ROUNDTABLE

The Government-University-Industry Research Roundtable is sponsored by the National Academy of Sciences, National Academy of Engineering, and Institute of Medicine. The Research Roundtable was created in 1984 to provide a forum where scientists, engineers, administrators, and policymakers from government, university, and industry can meet on an ongoing basis to explore ways to improve the productivity of the nation's research enterprise. The object is to try to understand issues, to inject imaginative thought into the system, and to provide a setting for discussion and the seeking of common ground. The Roundtable does not make recommendations, nor offer specific advice. It does develop options and bring all interested parties together. The uniqueness of the Roundtable is in the breadth of its membership and in the continuity with which it can address issues.

Library of Congress Catalog Card Number 89-63536

International Standard Book Number 0-309-04175-9

Printed in the United States of America

WORKING GROUP
ON
THE ACADEMIC RESEARCH ENTERPRISE

ERICH BLOCH, *(Chairman)*, Director, National Science Foundation
WILLIAM H. DANFORTH, *(Vice-Chairman)*, Chancellor, Washington University
KATHERINE L. BICK, Deputy Director for Extramural Research,
National Institutes of Health
JOEL S. BIRNBAUM, Vice President and General Manager,
Information Architecture Group, Hewlett-Packard Company
HAROLD H. HALL, Vice President (Ret.), XEROX Corporation
BARRY MUNITZ, Chairman and CEO, United Financial Group, Inc.
T. ALEXANDER POND, Executive Vice President and Chief Academic Officer,
Rutgers University
RUDI SCHMID, Associate Dean of International Relations,
School of Medicine, University of California
HAROLD T. SHAPIRO, President, Princeton University
LARRY L. SMARR, Director, National Center for Supercomputing Applications,
University of Illinois
ROBERT L. SPROULL, President Emeritus, University of Rochester
S. FREDERICK STARR, President, Oberlin College
LINDA S. WILSON, President, Radcliffe College
MARK S. WRIGHTON, Chairman, Department of Chemistry,
Massachusetts Institute of Technology
HARRIET ZUCKERMAN, Professor of Sociology, Columbia University

FRANK CARRUBBA, *(Associate)*, Director, Hewlett-Packard Laboratories
JOHN H. MOORE, *(Associate)*, Deputy Director, National Science Foundation

DALE R. CORSON, *(Advisor)* , President Emeritus, Cornell University

Staff

DON I. PHILLIPS, Executive Director,
Government-University-Industry Research Roundtable
JOHN P. CAMPBELL, Senior Program Officer,
Working Group on the Academic Research Enterprise
JAMES SINGER, Editorial Consultant
EVAN M. BERMAN, Research Consultant
SUSAN TAWFIK, Senior Secretary

ACKNOWLEDGEMENTS

The interest and support of the National Academy of Sciences, the National Academy of Engineering, and the Institute of Medicine, and of their presidents, for the wide-ranging deliberations of the Government-University-Industry Research Roundtable have made possible a thoughtful exploration of difficult issues affecting the research enterprise. The Roundtable Council and its chairman, James D. Ebert, encouraged a far-ranging examination of the characteristics of the enterprise and provided useful input throughout. The able staff of the Roundtable, especially Don I. Phillips, its executive director, and John P. Campbell, the project director for this effort, deserve special thanks for the difficult task of rendering a complex body of ideas and data into a readable and thought-provoking document.

Special thanks go to the staff of the National Science Foundation who have contributed to this effort, particularly the Division of Policy Research and Analysis, whose analysts collected, analyzed, and summarized the data on trends.

But the document is first and foremost the product of the Working Group on the Academic Research Enterprise. We hope we have assembled a coherent picture of the status of one of America's most valuable resources--its academic research enterprise.

PREFACE

The Government-University-Industry Research Roundtable was organized in 1984 under the aegis of the National Academy of Sciences, the National Academy of Engineering, and the Institute of Medicine. It is governed by a Council of 25 distinguished scientists, engineers, administrators, and policy makers from government, universities, and industry. Its purpose is to create a national forum to air the issues that affect the nation's research enterprise, inject imaginative thought into understanding the issues, and explore strategies and options for improving the future of U.S. scientific research. In short, the Roundtable brings together interested parties and develops options; it does not take sides, make recommendations, or offer specific advice.

In 1987, the Roundtable Council inaugurated a comprehensive review of the U.S. academic research enterprise. This effort was in response to concerns raised by the universities themselves, their research sponsors, and the general public. Among many concerns were the changing nature of science and engineering research, declines in the college-age population, the increasing financial and human resource requirements for carrying out research, and the growing expectations placed on the academic research enterprise. These concerns raised questions regarding the role of universities and colleges within the overall U.S. research system, the nation's ability to support the academic research enterprise, the management of universities and colleges, and the responsibilities of research sponsors.

The Council assigned this review to a Working Group of government officials, corporate executives, university administrators, and scientists. The charge to the Working Group was:

- Examine current trends in the university research enterprise.
- Predict the impact of the trends on the future of the enterprise.
- Determine the options for the future of the enterprise.
- Explore national strategies for meeting the challenges of the future.

The Working Group divided the project into two phases. Phase one would analyze the status, trends, and issues affecting academic research in science and technology, and examine the implications growth in these fields holds for the larger academic enterprise. During phase two, the Working Group will select for further analysis topics identified in phase one, and identify alternative options for the future of the enterprise and criteria for choosing among the alternatives.

In setting forth an analytic process, the Working Group took special note of the fact that science and technology comprise only two components in the full range of academic scholarship. Combined, however, they represent a large and discrete percentage of national financial support for academic research. Other components of academic scholarship--the arts and humanities, for example--also merit analysis. Their absence from this study, however, should not be construed as a statement of academic or public policy priority. They have meaningfully different cultures and requirements, and deserve independent inquiry beyond the capability of this Working Group.

This is a discussion paper describing the Working Group's progress in analyzing the status of scientific and technological research in academic settings and identifying issues central to its future. It is a working document, integrating the experiential knowledge of group members with quantitative analyses of available data. It should be noted at the outset that the quantitative information presented in this discussion paper primarily describes inputs to the academic research enterprise, such as financial and human resources. While some output measures have been developed--using publication and citation rates, patents, or departmental rankings--they require further methodological refinement before they can be meaningfully incorporated into analyses of academic research. Reliable data on long-term trends in academic research quality, productivity, or efficiency do not exist.

The purpose of this paper is to stimulate policy discussions--especially among individuals and organizations who have a direct role in funding or performing academic research. In the near future, the Working Group will hold a series of conferences for university, congressional, federal and state governmental, and industry officials, as well as academic scientists and engineers, to discuss options and alternative scenarios for sustaining the quality of academic research during the 1990s and into the next century. In preparing for those conferences, the Working Group invites candid responses to this paper; additional perspectives will enhance understanding of the issues and sharpen insights into the underlying influences on the academic research enterprise.

The paper has two parts. Part One analyzes the status of the current research enterprise, emerging trends affecting it, and major issues to be addressed regarding its future. Part Two provides an overview of the academic research enterprise, describing long-term trends in financial and human resources.

TABLE OF CONTENTS

PART ONE: STATUS, TRENDS, AND ISSUES

PART TWO: OVERVIEW OF THE ACADEMIC RESEARCH ENTERPRISE

LIST OF ILLUSTRATIONS

PART ONE: STATUS, TRENDS, AND ISSUES

PART TWO: OVERVIEW OF THE ACADEMIC RESEARCH ENTERPRISE

SCIENCE AND TECHNOLOGY IN THE ACADEMIC ENTERPRISE

PART ONE:

STATUS, TRENDS, AND ISSUES

INTRODUCTION

In the decades since World War Two, the United States has developed a unique research enterprise. Unlike most other industrialized nations, which developed basic research capacity primarily in government and industry laboratories, the United States expanded basic research within its universities as an adjunct to graduate education.[1] This coupling of functions has led to extraordinary success in the sciences and engineering. Two generally recognized factors have contributed to that success and the continued world-wide pre-eminence of the U.S. university-based research system: First, financial support for academic research and graduate education by federal and state governments, philanthropies, industries, and universities has significantly expanded and diversified the research enterprise; second, special reliance on the apprenticeship model of integrating advanced instruction and research has trained the nation's scientists on real research problems, not hypothetical exercises.

Coincident with this success, however, much is changing--in both the universities and the forces that influence and support them. As new pressures now urge expansion of the enterprise, for example, student enrollments and faculty positions are holding steady. As exciting research opportunities proliferate, the costs of pursuing them has grown sharply. As a large number of academic research faculty hired in the 1960s and early 1970s will begin to retire in the 1990s, the number of students planning careers in academic research may be inadequate to replace those faculty. In addition, research facilities built during previous decades need attention; they require repair and renovation and, in some cases, replacement. Scientific instrumentation, increasingly important to new opportunities on the research agenda, requires continual upgrading. Finally, the topics of scientific investigation, the ways in which academic scholars pursue research opportunities, and the role of university administrators are challenged by increasingly complex social and political demands.

Coupled with shifting economic, socio-demographic, and political climates in both U.S. and world society, these trends create a much different context for the academic research enterprise than the one that characterized its period of greatest expansion following World War Two. In the years ahead, these trends frame three major challenges to policy-makers in the enterprise and the nation:

- To maintain the overall quality of the nation's universities and their academic research, in an increasingly diversified enterprise with financial constraints.
- To ensure sufficient scientific and technical human resources.
- To enhance the nation's ability to address new scientific and technological opportunities and concomitant societal demands.

Achieving consensus to respond to these challenges will not be easy. Views will differ and vigorous debate is likely as all who hold a stake in the enterprise wrestle with these difficult and complex challenges. As it responds to them, the United States should re-examine its methods of conducting and financing research to assess whether the enterprise should be modified or restructured, and to determine how best to re-energize the imperatives of its mission and maintain the pre-eminence of its institutions.

Part One provides a framework for debating these issues and developing a consensus on the actions required. Three topics are addressed:

- The status of the current enterprise of academic research in science and technology.
- The emerging trends that affect the academic research enterprise.
- The major issues that will face research sponsors, university administrators, and academic scientists and engineers in the 1990s and into the 21st Century.

STATUS OF THE ENTERPRISE

The concept that scholarship and advanced research training should be conducted jointly in institutions of higher learning has been a major tenet of most leading U.S. universities for more than a century. This tenet, imported with significant modifications from the great European universities, not only promotes a university-based model for the development of new knowledge, but also stimulates faculty to gain the forefront in contemporary science. As a result, this dual emphasis on new knowledge and pedagogy has established a unique inter-dependence between education--including advanced research training--and research in the United States; universities educate new generations of teachers, researchers, and other professionals, as well as produce fundamental knowledge for science and social, economic, and cultural development.

Figure 1-1: Leading U.S. Research Universities Based on the Number of Distinguished Faculty, 1906*

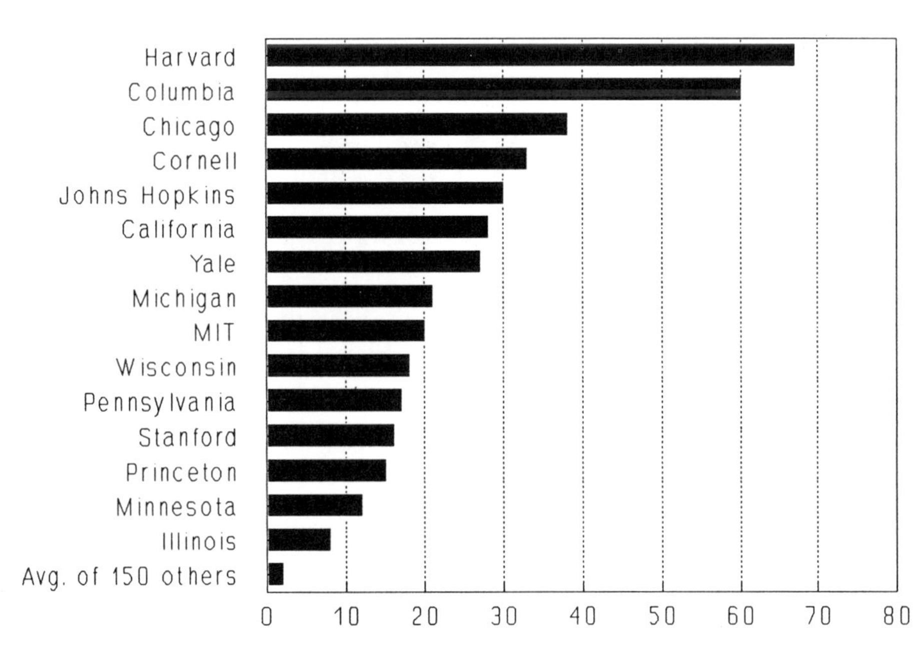

* From J.M. Cattell, *Science,* 1906

By the end of the 19th Century, about 15 U.S. colleges and universities had undergraduate enrollments of sufficient size to organize their faculties into specialized departments (Figure 1-1). From the beginning, external funding was critical for university-based research but generally limited to small endowments and government appropriations for agricultural experiment stations.[2]

After World War One and throughout the 1920s, the academic research enterprise grew significantly through two sources: Increased numbers of faculty due to rising undergraduate enrollments (Figure 1-2), and the emergence of external sponsors for research. These sponsors were, principally, philanthropic foundations, which awarded block grants to major private universities, and industries, which underwrote programmatic grants in their areas of commercial interest. Direct federal support remained small. Much of the private funding, however, was short-lived. The Great Depression of the 1930s significantly reduced private sector support, and academic research entered a decade of doldrums that did not end until the onset of World War Two.

Figure 1-2: Growth of U.S. Higher Education and Major Socioeconomic Influences, 1900-1988*

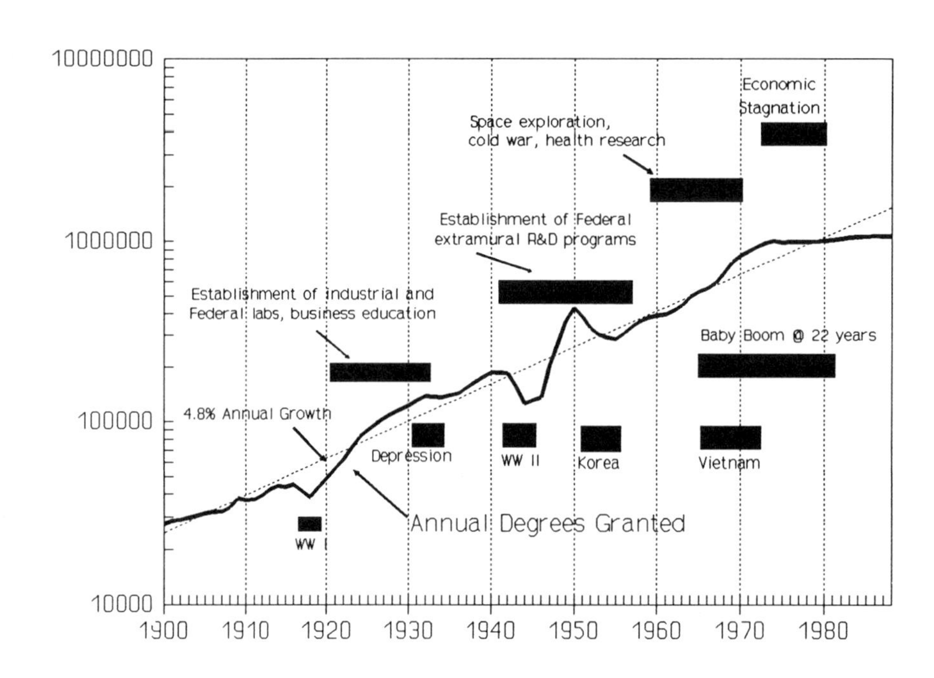

* Measured by bachelors and first professional degrees. Source: National Science Foundation.

The Second World War was a turning point. Academic scientists greatly assisted the national government during the war and, with the war's end, national policy-makers perceived a direct link between the seminal role basic research had played in ending the hostilities and the need to develop creative solutions to major social problems.[3]

After the war, federal policy-makers acted to put in place an enterprise that could direct the contributions of research to national needs. They made two historic decisions that fundamentally re-shaped the academic research enterprise: First, the federal government assumed primary responsibility for the quality and quantity of basic research in the United States and, second, the government identified the universities as the primary locus for the increased basic research activity.

The first decision, in effect, established university reliance on the federal government for financial support. In 1960, the President's Science Advisory Committee explained the rationale for this decision:

Whether the quantity and quality of basic research and graduate education in the United States will be adequate or inadequate depends primarily upon the government of the United States. From this responsibility the Federal Government has no escape. Either it will find the policies--and the resources--which permit our universities to flourish and their duties to be adequately discharged--or no one will.[4]

The second decision meant, in operational terms, that U.S. basic research and graduate education would be carried out as joint university activities. A 1964 report of the National Academy of Sciences described this teaching-research relationship as fundamental to the success of U.S. science. The report recommended against placing basic research in non-university laboratories and strongly opposed hiring distinguished scientists for non-teaching university research positions:

Graduate education can be of highest quality only if it is conducted as a part of the research process itself. The research must not be in the form of mock problems; it must be a part of the exploration of the unknown, with all the uncertainties and challenges that go with it. By the same token, research can remain truly a quest, with freedom to follow unexpected lines, if the tentative conclusions of recent scientific research are tested in the interplay of advanced teaching.[5]

Over the past three decades, the two decisions have been tested by strain and tension--direct results of changing patterns in financial support, employment of academic personnel, and student enrollments. Each decade has presented policy-makers with a unique set of challenges, problems, and opportunities.

1958 to 1968: Expansion

Between 1958 and 1968, the major challenge was the management of growth. By 1968, academic institutions conducted half of the nation's basic research, up from 30 percent in 1958 (Figure 1-3).* In addition, academic research more than doubled its share of the nation's economy, growing from 0.10 percent to nearly 0.25 percent of the gross national product during the same period (Figure 1-4).

With rapid growth in funds for basic research, total academic research and development expenditures more than tripled, from under $2 billion (in 1988-constant dollars) in 1958 to nearly $7 billion in 1968 (Figure 1-5).** The greatest growth rates occurred in the life and social and behavioral sciences (Figure 1-6). Academic research personnel in public universities that award doctoral degrees grew from 13,000 to 23,000.[7] In private doctoral universities, the growth was similar, from 12,000 researchers in 1958 to

---

*Figures 1-3 through 1-20 begin on page 1-11.

**Unless otherwise noted, all data regarding academic research include the following broad academic fields: life and health sciences, physical sciences, environmental sciences, engineering, mathematics, computer sciences, and social and behavioral sciences.

23,000 in 1968 (Figure 1-7).[6] With increased funding, average combined operating and capital expenditures per researcher rose from $85,000 to $170,000 (Figure 1-8).

Two driving forces produced this extraordinary expansion: One, a substantial increase in the number of faculty (Figure 1-9) due to surging university and college enrollments and, two, urgent and substantial increases in federal investment in academic research--fueled by anxiety over the national security, desire for international leadership, and recognition of general domestic problems.

But if the Cold War, Sputnik, and concern over cancer and heart disease provided the impetus, the burgeoning U.S. economy provided the means. From 1958 to 1968, annual federal contributions to academic research increased five-fold, from $1 billion (1988 dollars) to $5 billion (Figures 1-10 and 1-11).

The federal share of public doctoral universities' research funds increased from 53 percent to nearly 75 percent; for private doctoral universities, the federal share increased from 66 percent to 82 percent (Figures 1-12 and 1-13). The preponderance of growth in federal R&D spending occurred in non-defense agencies (Figure 1-16).

Simultaneously, a rapidly growing job market for college graduates and the maturation of the post-war baby boom doubled the size of the U.S. higher education system, rapidly expanding the institutional base for academic science and technology. Between 1958 and 1968, total higher education enrollments rose from 3 million to more than 7 million, as 2-year colleges firmly took their place in the education system. But the universities that offered doctoral programs grew also. Enrollments in public doctoral universities, for example, doubled from 800,000 to 1.9 million during the decade, while private doctoral university enrollments grew from 440,000 to 650,000 (Figure 1-17). The increase is more striking for advanced degrees awarded during this period. Annual Ph.D. degrees granted in the sciences and engineering from public institutions nearly tripled, rising from 3,300 to 9,000 per year, and those granted by private institutions doubled, from 2,500 to 5,300 (Figure 1-18).

1968 to 1978: Steady-State

In contrast to the previous decade, the major challenge for the period between 1968 and 1978 became managing steady-state funding for the academic research enterprise. The decade began with an expanding guns-and-butter federal budgetary policy and ended with national belt tightening. Accounting for inflation, total academic research expenditures for the decade showed no real growth, fluctuating around $7 billion (1988 dollars); as a share of the gross national product, academic research declined from 0.25 percent to 0.21 percent (Figures 1-4). When inflation is accounted for, annual federal contributions to academic research declined from $5 billion in 1968 to $4.7 billion in 1974, then increased again to $5 billion in 1978 (Figures 1-10 and 1-11).

During the period, the federal share of public doctoral universities' research funds decreased from 75 percent to 60 percent; for private doctoral universities, the federal share decreased from 82 percent to 77 percent (Figures 1-12 and 1-13). While the number of research personnel in public doctoral universities continued to grow, from 23,000 to 32,000, for private doctoral universities, the number declined from 23,000 to under 20,000 (Figure 1-7). With flat funding, average research expenditures (operating and capital) per academic researcher hovered around $160,000 (1988 dollars) (Figure 1-8).

A major legacy of the leveling off of federal research funding was doubt raised about the continued federal stewardship of basic research in the United States.[7] Many factors underlay the changing pattern of federal support, including rising general inflation, economic recession, the end of the manned moon mission, the Vietnam War, increased budgetary competition from other federal programs, and a re-assessment, by both government and universities, of the relationship between the federal government and the universities. In the view of some policy-makers, the institution building objective had been achieved by the 1970s, perhaps even over-achieved, and attention should be turned to the management of the expanded enterprise. Others in the academic research community feared that a long-term steady-state in federal support would reduce both the size and quality of the enterprise. Policy debates focused on cutbacks in federal support--primarily for student fellowships, facilities, and equipment--and increasingly restrictive regulations for monitoring the expenditure of federal research dollars by universities.

The institutional base for academic research also approached steady-state. While enrollments continued to grow rapidly in comprehensive universities and 2-year colleges, enrollments stabilized in doctoral research universities by 1973. In the public doctoral universities, total enrollments reached 2.5 million; private doctoral universities enrollments slowly increased to 700,000 (Figure 1-17).

With an approaching steady-state in faculty positions and uncertain federal financial support for research, the production of Ph.D. degrees in the sciences and engineering began to drop. Annual Ph.D. degrees granted in the sciences and engineering from public institutions peaked in 1973 at 12,500, then declined 10 percent to 11,100 by decade end. Production in the private institutions fared worse, declining 18 percent from a high of 6,500 in 1973 to 5,300 in 1978 (Figure 1-18).

Together, the uncertainties of funding and university enrollments generated doubts about continued federal commitment to basic research and the ability of universities to remain its primary locus.

1978 to 1988: Diversification

The years from 1978 to 1988 saw a dramatic diversification in the academic research enterprise. The fears expressed in the previous decade that the enterprise would contract did not prove out. Rather, a new infusion of research dollars spurred a broader range of academic institutions to develop research capacity and participate in the enterprise. (See Figures 1-14 and 1-15.) Competition for faculty and research support increased; so did competition for students as the enrollment inertia of the previous decade continued.

In inflation-adjusted dollars, support for academic research nearly doubled, rising to more than $13 billion in 1988 from less than $8 billion (1988 dollars) in 1978 (Figure 1-5), and reached an all-time high 0.27 percent of the gross national product, up from 0.21 percent in 1978 (Figure 1-4). Annual federal contributions increased from $5 billion to $8 billion (Figures 1-10 and 1-11). The number of research personnel--faculty and non-faculty--in public doctoral universities grew from 32,000 to 40,000; in private doctoral universities, personnel increased from below 20,000 to more than 22,000 (Figure 1-7). With increased funding, average expenditures per academic investigator rose from $160,000 to $220,000 per year (Figure 1-8).

The sources of funding support also diversified, adding fuel to the questions and doubts about continued federal responsibility for academic research. While federal funding grew over the decade, non-federal funding grew even more dramatically. From 1978 to 1988, the federal share of academic research support declined from 66 percent to 60 percent (Figures 1-10 and 1-11). Among private doctoral universities, the federal share decreased from 75 percent to 73 percent, while among public doctoral universities, it dropped from 60 percent to 53 percent (Figures 1-12 and 1-13).

In contrast to the decline in federal share, university-generated research funds grew from 12 percent to 18 percent.[8] The most significant factor in this trend in university funding was the willingness of public universities--especially those aspiring institutions who were just beginning to develop a research base--to allocate their own resources to cover a significant share of the indirect costs associated with externally sponsored research.[9]

Industry also took a larger role, nearly doubling its slice of academic research funding from 3.7 percent to 6.5 percent. The industry support tends to be concentrated in certain research areas and certain institutions; in these instances, it is becoming an influential force.

Although the over-all state government share of academic research funds held steady at 8 percent, several state governments dramatically increased their individual contributions to academic research.[10] While much of this support focuses on applied research to meet the needs of local industries, it has the potential for developing future basic research capacity at scores of campuses where earlier it scarcely existed.

The diversification in sources of research support reflected significant and fundamental changes that were occurring elsewhere in the research enterprise--the decentralization of scientific research from a small number of academic centers that dominated the enterprise before World War Two to a wider array of institutions, and, in the political arena, a sudden determination by civic leaders in many areas of the country to enhance the research capacity of local universities for economic development purposes. Premier research universities, of course, continued to dominate most fields of science, but infusions of state funds enabled aspiring public institutions to achieve real annual growth rates in research funds in excess of nearly 5.5 percent--higher than that of the top-20 research universities.[11]

While the academic research enterprise continued to expand, however, the number of students stabilized at about 2.7 million per year for public universities and 750,000 per year for private universities (Figure 1-17). The Ph.D. degrees granted in science and engineering by public institutions increased to 13,600 in 1988, up from 11,200 in 1978. Ph.D. degrees granted by private institutions grew from 5,300 to 6,600 per year. (Figure 1-18). This renewed growth in Ph.D. production is primarily due to a rising enrollment of foreign students in the natural sciences and engineering (Figures 1-19 and 1-20).[12]

If it was not clear earlier, it became so by the end of the 1979-1988 decade: The historic relationship between university research and graduate education was under stress from virtual steady-states in university enrollments and the over-all production of new doctoral researchers, on the one hand, and mounting pressure to expand basic research activities, with or without instructional components, on the other. With the over-all ratio of students to faculty remaining constant over the past decade, expansion occurred in part by creating extra-departmental research centers and institutes and hiring non-teaching researchers to operate them.[13] While graduate education in the United States continues to include significant research components, what appears to have changed is the extent to which expanding academic research programs include instructional components.

With regard to undergraduate education, all of these factors combine to provide disincentives to teaching. The increasing scale and organizational complexity of much new academic research activity, a faculty salary system that increasingly rewards research accomplishments, and federal policies which favor research over educational programs further exacerbate this situation. Some aspiring research universities, in response, have developed two-tier faculty systems--one tier for non-teaching research "superstars" and the other for teaching faculty.

Forces for Expansion and Diversification

The expansion of resources for the academic research enterprise during the past decade was spurred by powerful new expectations for science and technology--improvement in international competitiveness, aggressive state and local economic development, and growing research competition among the universities and colleges themselves. Unlike the expansion during the 1960s, which largely concentrated on institution building in the then-existing university research community, the current expansion is more the result of diversification--a continuing broadening in the number of institutions participating, increases in the number and types of organizations funding extra-mural research, and a broadening in the national research mission, particularly in support of such social problems as health, the environment, and economic competitiveness.

Concern for improving the nation's international competitiveness has generated expectations that universities, in partnership with industry, will provide scientific and technological breakthroughs in key commercial areas. At the federal level, for example, the country's dependence on the research enterprise takes on a new intensity as major international competitors' investments in research grow at a faster pace than ours--signaling an intensification in economic rivalries. But the federal government isn't alone in its renewed interest in academic research; industry also is demonstrating interest, a significant portion of which represents an increased reliance on universities for entree to basic research frontiers. New commercial technologies, in turn, generate and make possible the exploration of new basic research avenues. To achieve these mutual interests, industries are augmenting the research capacity of U.S. universities.

State and local officials increasingly urge their public universities to contribute to regional development through applied research and cooperation with resident industries; they recognize that local academic research is often a magnet, drawing high-tech industries and new jobs to an area. As the economic benefits of academic research catch public attention and imagination, political leaders press for a larger and geographically broader academic research enterprise. In addition, some federal research appropriations are earmarked for specific locations, often on a basis of economic development or local scientific research agendas.

Competition among universities also helps to drive the current expansion in research. The major universities are enlarging their research capacity to maintain their competitive standing. Aspiring research universities are under great pressure to develop research capacity; they are also at some financial risk, whether they opt to develop research capacity or not. If they seek to attract a prestigious scientific and engineering faculty, they must invest resources heavily in state-of-the-art research facilities and instrumentation; in a competitive academic labor market, even promising younger faculty members can now demand university resources for their research projects and time to establish their research careers before undertaking teaching duties. If, on the other hand, universities do not seek to expand their research capacity, they now jeopardize financial, political, and community support for their institutions.

Figure 1-3: Distribution of U.S. Basic Research Expenditures by Performer

(See Figure 2-20)*

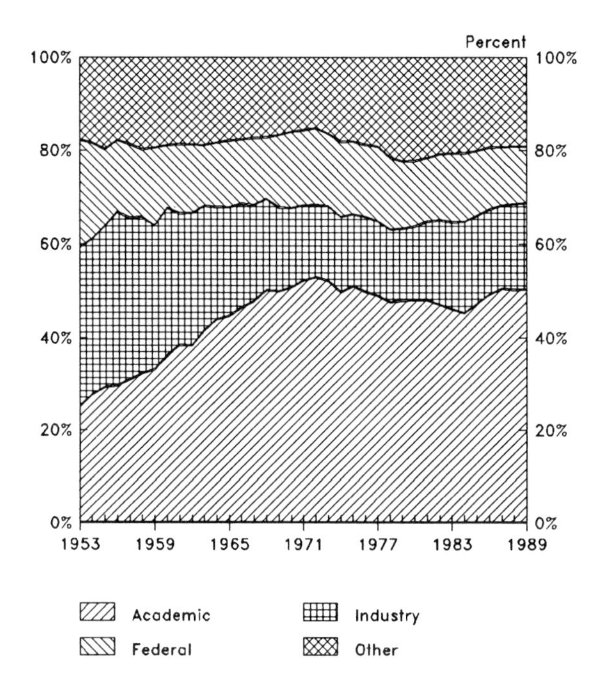

Figure 1-4: Total and Federal Academic R&D Funds as Percents of the Gross National Product

(See Figure 2-4)*

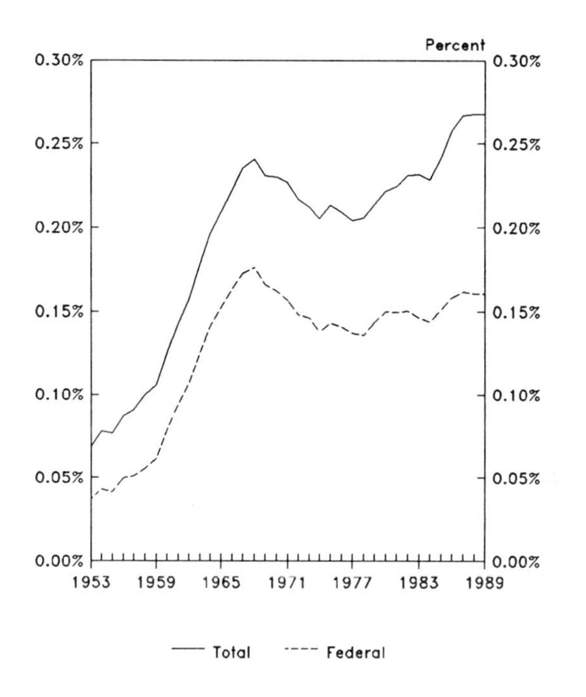

Figure 1-5: Academic R&D Expenditures By Type of R&D

(See Figure 2-23)*

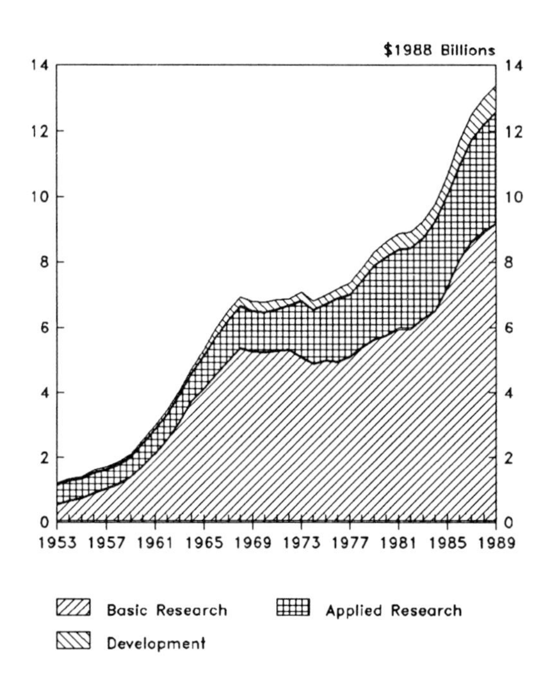

Figure 1-6: Distribution of Academic R&D Expenditures by Science and Engineering Field

(See Figure 2-30)*

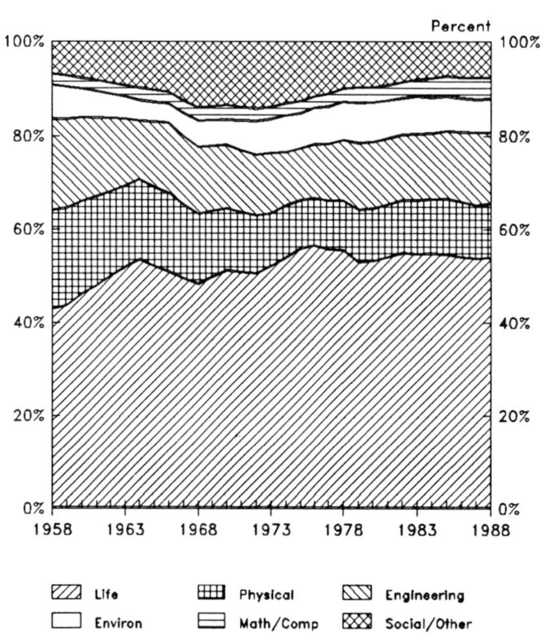

* See corresponding figure in Part Two for data sources and definition of terms.

Figure 1-7: Investigators (FTE) in Doctoral Institutions by Institution Governance

(See Figure 2-74)*

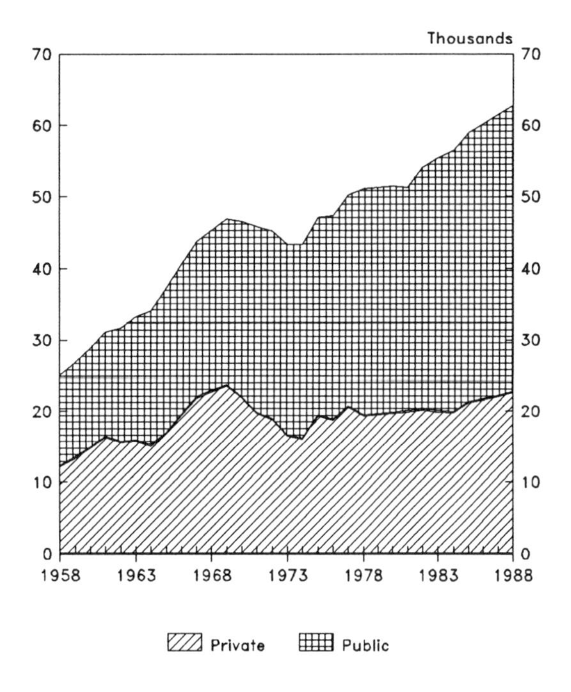

Figure 1-8: Academic R&D Expenditures per FTE Investigator by Type of Expenditure

(See Figure 2-45)*

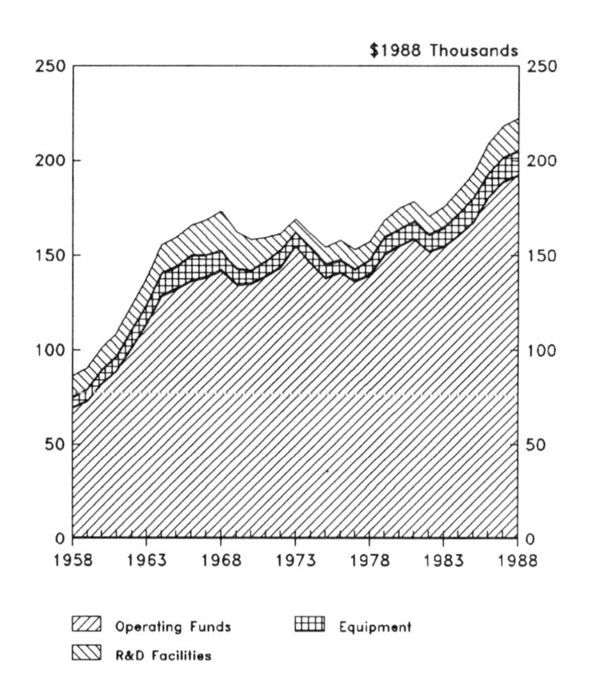

Figure 1-9: Academic Faculty by Institution Type

(See Figure 2-64)*

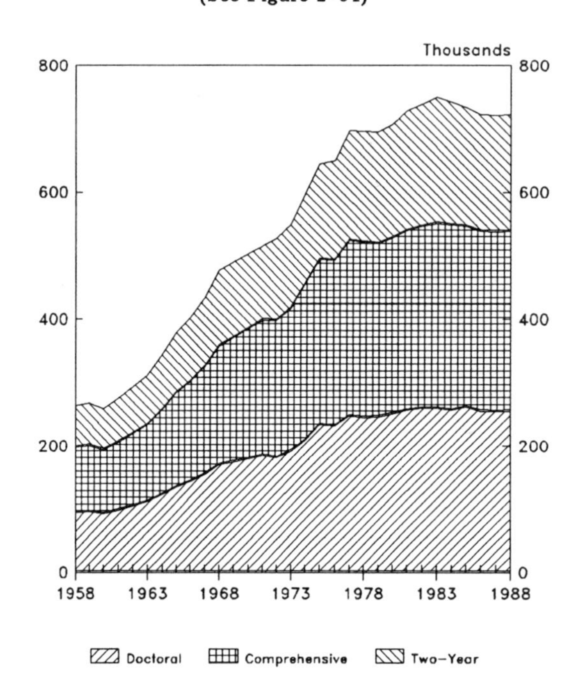

*See corresponding figure in Part Two for data sources and definitions.

Figure 1-10: Academic R&D Expenditures by Source

(See Figure 2-27)*

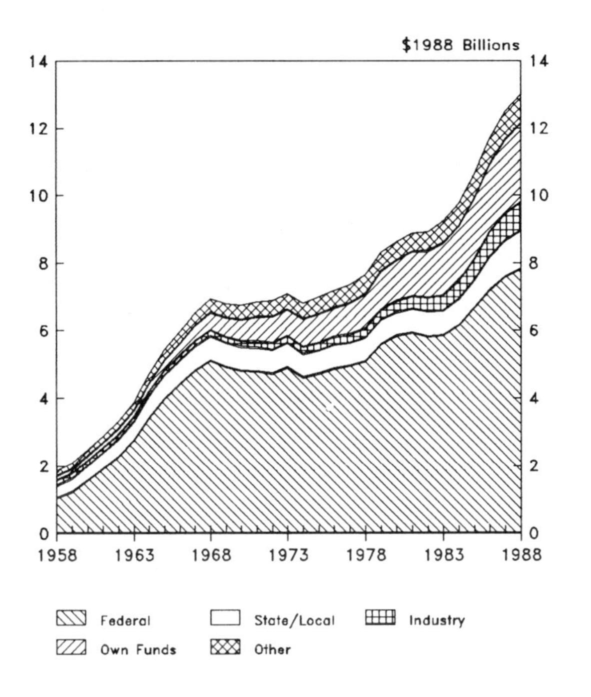

Figure 1-11: Distribution of Academic R&D Expenditures by Source

(See Figure 2-28)*

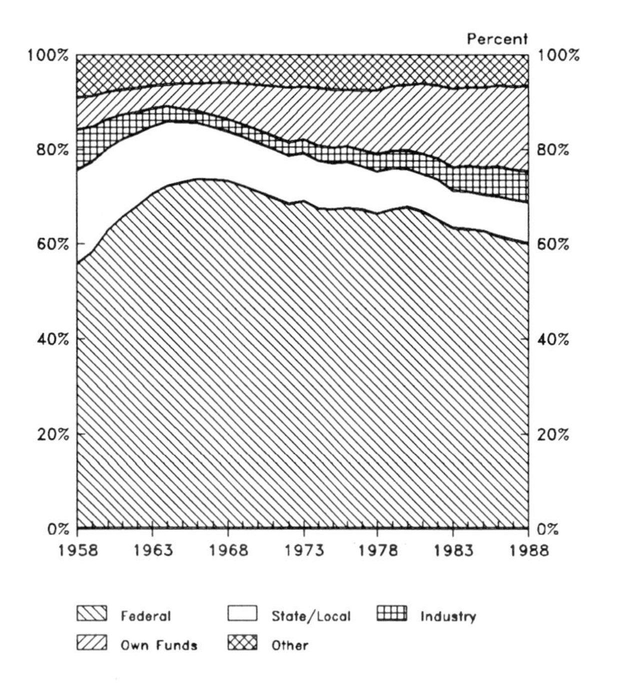

Figure 1-12: Distribution of Public Doctoral Institution R&D Revenues by Source of Funds

(See Figure 2-34)*

Figure 1-13: Distribution of Private Doctoral Institution R&D Revenues by Source of Funds

(See Figure 2-32)*

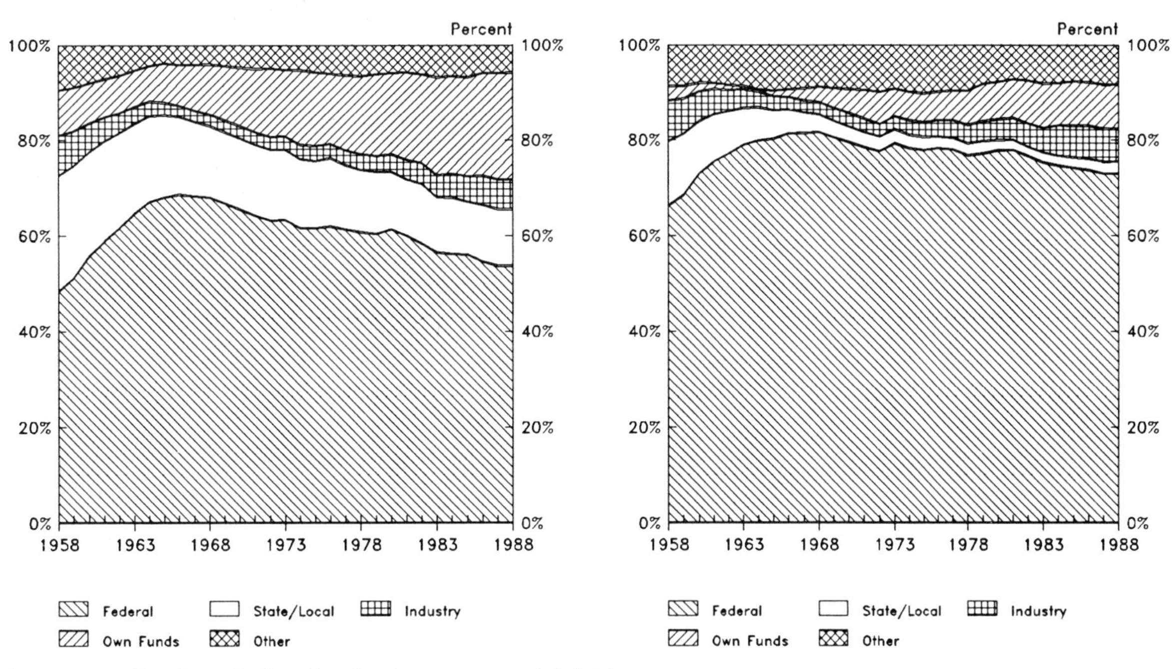

*See corresponding figure in Part Two for data sources and definitions.

Figure 1-14: R&D Expenditures among Doctoral Institutions*

Figure 1-15: Distribution of R&D Expenditures among Doctoral Institutions*

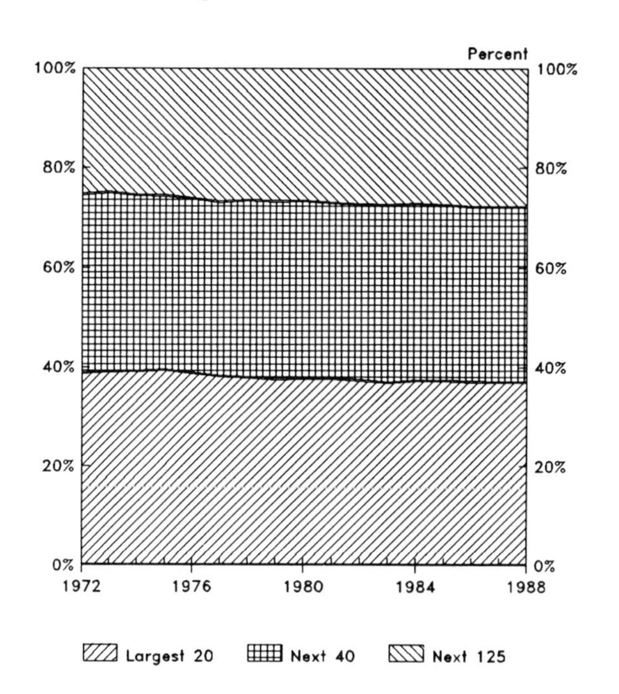

Figure 1-16: Distribution of Federal Academic R&D Funding by Federal Agency, 1945-1988**

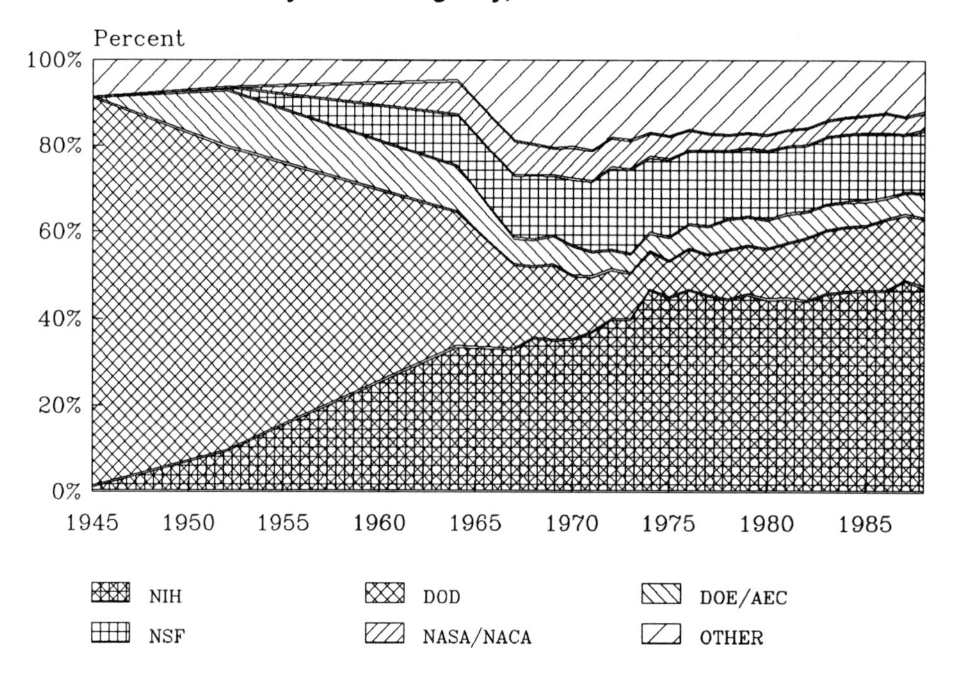

* Source: National Science Foundation. *Largest 20* includes the 20 doctoral institutions with the largest R&D expenditures. *Next 40* includes the next 40 institutions with largest R&D expenditures, and *Next 125* includes all other doctoral institutions.

**Source: National Science Foundation.

Figure 1-17: Enrollment in Academic Institutions by Institution Type and Governance

(See Figure 2-78)*

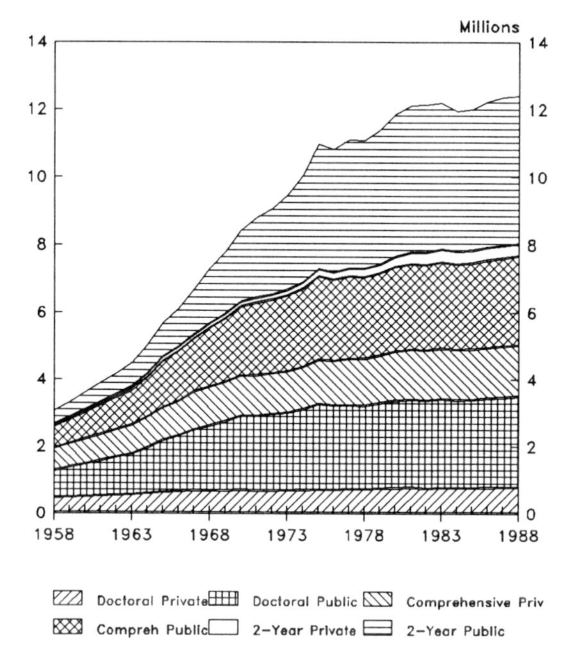

Figure 1-18: Ph.D. Degrees Awarded in Science and Engineering by Institution Governance

(See Figure 2-94)*

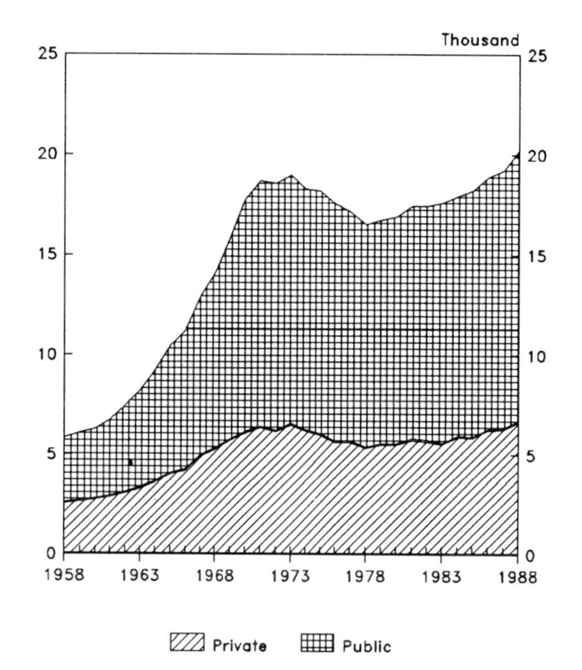

Figure 1-19: Ph.D. Degrees Awarded in Engineering by Citizenship

(See Figure 2-106)*

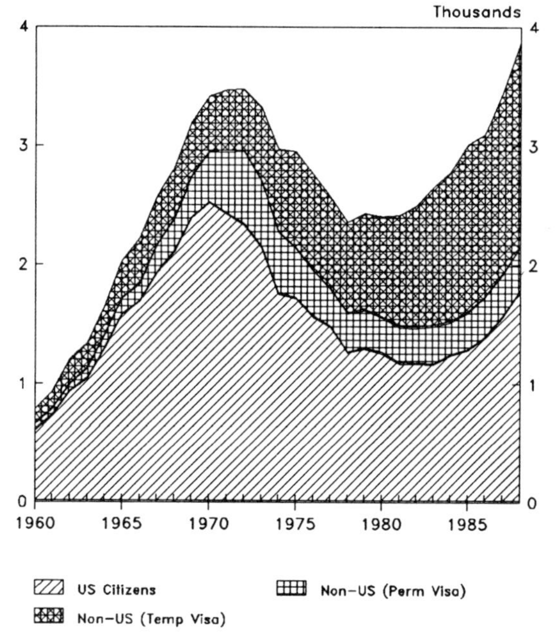

Figure 1-20: Ph.D. Degrees Awarded in Natural Sciences by Citizenship

(See Figure 2-104)*

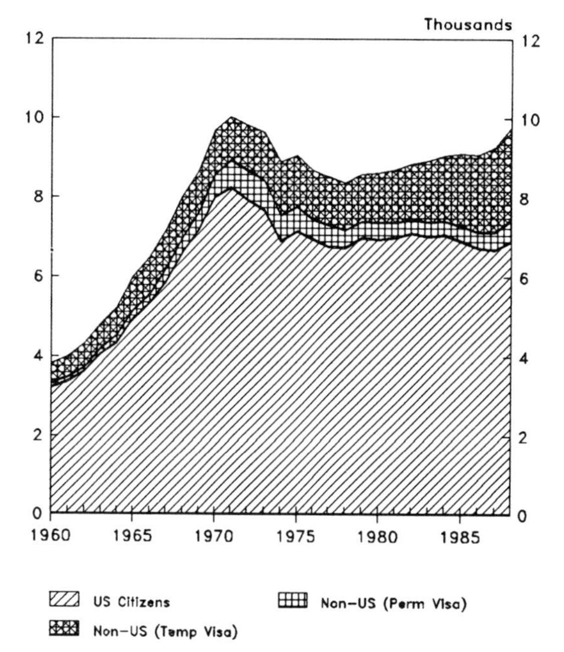

*See corresponding figure in Part Two for data sources and definitions.

EMERGING TRENDS

The ability of universities to broaden their missions and play a larger role in the nation's research enterprise will depend on the resolution of three sources of tension, each pulling at the fabric of the enterprise. The first strain on the enterprise is slow adaptation to an increasingly complex research and educational environment; the organization, culture, and resources of academic institutions and their research sponsors constrain their response to new demands and opportunities. The second source of stress on the enterprise is the replacement of retiring high-quality research personnel during the next decade; it may not be possible, given the current production level of research scientists and engineers. The third source emanates from the need to sustain the quality of current research institutions and programs, which is increasingly expensive to do and--in an era of severely constrained fiscal resources--increasingly difficult.

The Research Environment

The environment in which the academic research community must function will increase in complexity. National and international economic, political, and social cross-currents influence the priorities, topics, and contexts of scientific investigation. These influences are combining to challenge the traditional way scholars and their host institutions operate and relate to each other. Furthermore, many new scientific and technological opportunities require more flexible, cross-disciplinary relationships both within and among universities, industries, and governments.

There are many factors at work here. First, important and exciting advances in fundamental science are occurring are creating more complex questions on the research frontier and many of the questions are more frequently in multi-disciplinary settings at the interface between disciplines. Furthermore, some traditional fields, such as molecular biology and microelectronics, are merging with other fields or being redefined.

Second, as product life cycles become shorter, advances in fundamental knowledge become more relevant to technology development. As a result, industries, universities, and financial institutions are developing sophisticated relationships that include a multiplicity of formal and informal structures. Some faculty members, for example, are assuming entrepreneurial roles, including developing relationships with non-academic organizations to pursue the commercial development of their research.

Third, international cooperation is intensifying in many scientific and engineering fields. The growing research capabilities of other nations provide new opportunities for collaboration--especially in astronomy, oceanography, and high-energy physics--that require large capital investments. International cooperation is also required for research on such problems as global climate change, ozone depletion, and acid rain.

New technologies increasingly shape the scholarly agenda in the sciences and engineering. State-of-the-art instrumentation allows for experiments requiring heretofore un-achievable precision and scale. New generations of computers make possible large-scale data analysis and provide the mechanism for rapidly transferring and sharing information among institutions, organizations, and nations.

News of new processes and products of scientific research reach an ever-wider U.S. audience. To the extent that popularization contributes to public understanding of science, it enhances political support. But it also brings greater societal scrutiny to the research enterprise. There is, for example, growing public pressure on federal regulatory and grant-making agencies to control the use of toxic substances and radioisotopes, and experiments involving animals. In addition, societal intervention in the research agenda is increasingly exercised through the courts, notably in environmental protection, radiation and carcinogen disposal, and the release of genetically engineered material. In addition to increasing regulatory complexity in some fields, the lack of regulations in other fields is also a problem--often forcing researchers to curtail or abandon lines of inquiry in areas such as biotechnology.

The most pronounced recent trend is state and local regulation of research. A few state, county, and city governments have begun to influence the conduct of local university research through controls on the type and location of university facilities and on research protocols, such as the use and care of test animals and the use of genetically altered organisms. Should this trend become more widespread, investigators and their host institutions would have to adapt to a changing array of costly reporting requirements, safeguards, controls, and regulatory supervision.

Universities and research sponsors face difficulty in rapidly adapting to a changing research environment. In response to the changing research environment, some members of the academic enterprise are testing innovative strategies for organizing, conducting, managing, and financing research. Rapid adaptation to new demands and opportunities in the research area, however, is slowed by many factors--including tradition, inertia, the competition for university resources, the demands of the university's educational mission, and the aging of faculty--impinging on the current organization, culture, and resources of university-based scholars and their funding agencies.

There is growing debate within universities over the ability of the current disciplinary and governance structures to respond adequately to the expanding research agenda, as well as to find an appropriate balance of commitments to scholarship, education, and public service. New research opportunities often require more flexible budgeting and assignment of research faculty, inter-disciplinary approaches, expansion of non-faculty research personnel, extra-departmental initiatives, and allowance for faculty entrepreneurial activity. Furthermore, larger-scale multi-disciplinary research efforts require hierarchical management and more centralized governance structures for rapidly making strategic decisions and for inter-departmental planning. In addition, the intense regulatory environment in many areas of research requires active participation by the institution's administration in deciding faculty research topics and protocols, as well as in serving as a necessary buffer against unwarranted outside interference.

On the other hand, the present university disciplinary structure has proved adaptable to new research opportunities and, more importantly, provides a necessary, albeit cumbersome, system for quality control through peer review. Young faculty, who are strongly trained in disciplines, enter a reward system that favors a single-discipline setting to establish professional credentials. Moreover, the traditional collegial culture of universities, including the faculty tenure system, provides an atmosphere essential to fostering the creative process and maintaining academic proficiency.

For the external sponsors of academic research, the topics and capital requirements of new research opportunities pose challenges to their decision-making and budgetary structures. Inter-disciplinary research opportunities generate pressure for federal funding mechanisms that cut across divisions within a given agency, and often across agencies. Collaborative ventures among government funding agencies are often limited by competing Congressional committee jurisdictions and federal agency bureaucracies, and conflicting procedures and legal restrictions. The active participation of state governments in funding research provokes demands for federal-state consultation and cooperation in funding decisions. Among industries, collaborative ventures for supporting academic research are often constrained by anti-trust laws, competitive pressures, and trade secret and patent rights concerns.

Research Personnel

During the next decade, faculty retirements will increase demand for academic research personnel. Steady-state student enrollments during the past two decades have reduced the number of new faculty job openings. As a result, between 1973 and 1987, the percentage of academic scientists and engineers under 35 years of age fell from 27 to 12 percent.[14] This aging of the faculty indicates an increased number of faculty are slated for retirement in the foreseeable future. In some instances, however, the impact of these retirements may be eased temporarily by the end of mandatory-retirement policies and movement of non-tenure-track personnel into faculty positions. The risks of such solutions, however, are that they may dissuade students from choosing academic careers by reducing placement opportunities for new graduates.

Fewer numbers of U.S. students are now interested in or qualified for academic science and engineering careers. The number of baccalaureate degrees in science and engineering awarded to U.S. citizens has stabilized or declined in most fields. This situation results from the current decline in the college-age population and the steady rate at which 22-year olds attain such degrees. In the early 21st century, enrollments may slowly return to 1983 levels, riding an upswing in the number of 18- to 22-year olds. During the next several decades, however, assuming current enrollment rates, U.S. higher education enrollments will most likely not exceed current levels.[15] Nor is it likely that increased participation of women, minorities, and foreign students in undergraduate science and engineering programs will offset these general demographic declines.[16]

Since the mid 1960s, the rate at which students with natural science and engineering baccalaureate degrees from U.S. institutions went on to earn Ph.D.s has declined by half. This reduction has been especially apparent among U.S. males, a group that has historically been the mainstay for doctoral degrees. The recent growth in Ph.D. awards in several fields is due in part to greater participation by foreign students. In engineering, almost 60 percent of all doctorates are now awarded to foreign students, as are over a third of doctorates in mathematics and physics. Approximately half of all foreign students remain in the United States, making valuable contributions to the nation's economy, research, and education. However, the large numbers of foreign students involved and the likelihood that they will return in increasing numbers to take advantage of improved career opportunities in their homeland raises serious questions about the drain of much needed scientific knowledge and technical experience.[17] Increases in Ph.D. degrees in the biological sciences primarily result from the growing participation of U.S. females. Although the continuation rate for U.S. citizens into Ph.D. programs appears to be increasing, there is still concern that it will be inadequate for meeting academic labor demands in the next decade.

These trends in the potential supply of academic personnel, however, must be seen in the context of trends in education and training throughout U.S. society. The nation requires increasing supply of highly trained personnel in all economic sectors.

Financial Resources

Sustaining the quality of current research institutions and programs is increasingly expensive. An accelerating pace in the development of knowledge generates a proliferation of research opportunities. It is a self-reinforcing phenomenon: A theoretical or technological breakthrough--in any field, molecular biology, high-energy physics, or computer science--provokes demand for expensive new research. Increasing numbers of scientists and engineers, in pursuit of such exciting opportunities, propose sophisticated research designs, which often require additional laboratory space and equipment, and highly trained personnel. Universities and research sponsors, committed to maintaining their place at the frontier of scientific advance, are pressured to approve the proposed research.

High-quality research on the frontier of any discipline is increasingly capital intensive. In all sciences, the term "state-of-the-art" implies a technological sophistication of equipment and facilities that is increasingly costly, especially as dramatic technological advances accelerate the obsolescence of vast portions of existing equipment and facilities. This rapid pace in technological change is indicated by the fact that, in 1986, the median age of all academic research instrumentation classified as state-of-the-art was only 2 years old; in computer science, electrical engineering, chemistry, and environmental science, the median age was 1 year.[18] Other factors are also involved in equipment costs. One of the more important is the expense associated with keeping highly trained technicians on staff; another is a growing awareness of essentials for environmental and work-place safety, which inevitably drive up costs.

University research facilities, many built during the 1960s' boom years, need to be renovated or replaced. Recent surveys indicate that $3.4 billion is obligated nationally for construction of academic science and engineering facilities. University administrators estimate, however, that about $8.5 billion in necessary construction has been deferred. In repair and renovation alone, $777 million has been obligated for academic research facilities, but almost four times that amount has been deferred.[19] This, in effect, is an unfunded liability of nearly $3 billion and it continues to grow. This represents a potential danger to the long-term viability of these institutions.

The average compensation of an academic researcher has risen sharply in the last few years.[20] The reasons for this seem to be the result of two important factors: First, universities have to compete with industry for research personnel in several fields. Second, competition among universities for top research faculty fuels wage costs. In this regard, it should be noted that during the 1990s, wage pressures will likely continue to intensify because of the shortage of and demand for teaching Ph.D.s, particularly if an increase in student enrollments materializes. Growing demand by industry for Ph.D.s, driven by the complex technological base of the service, manufacturing, and agricultural sectors, will also fuel wage increases.

The United States has entered a period of constrained fiscal resources. In the nation's current economic circumstances, financing the perceived needs of the academic research enterprise will not be easily accomplished. Government policies during the next decade will be affected strongly by the large federal budget deficits and public resistance to raising taxes. State governments--many of which are confronting budgetary constraints--appear to be closely evaluating their needs and priorities, including the funding of academic research. In addition, industry-sponsored research may flatten or decrease, potentially exacerbated by corporate mergers and leveraged buy-outs. These pressures will intensify competition for available federal dollars and foster priority setting among federal programs. Academic research funding will not be immune from these processes.

The ability of many universities to generate significantly greater research funds through internal resources is likely to be limited. For public universities, for example, steady enrollments and state budget constraints may press the limits on state appropriations. For both private and public universities, constraints on tuition increases and additional philanthropic contributions may diminish their ability to maintain world leadership in research.

ISSUES FOR THE 1990s & BEYOND

The trends of the past decade converge to pose three major challenges to the academic research enterprise. As stated at the outset of this report, the challenges are: To maintain the over-all quality of the nation's universities and their academic research in an increasingly diversified, financially constrained environment; to ensure sufficient scientific and technical human resources to meet the nation's research mission; and to enhance the nation's ability to address new scientific and technological opportunities and concomitant societal demands.

To meet these challenges, it will be necessary to confront and resolve a number of complex, often inter-connected issues that affect the current status of the academic research enterprise. To begin that process, the Working Group here sets out what it believes to be the most critical issues confronting the enterprise. They are organized in five categories: Role of universities, the organization and management of universities, conduct of research and transfer of knowledge, education of scientists and engineers, and funding of academic research. The Working Group believes all are relevant to each stakeholder in the enterprise.

Role of Universities

What is the optimal role of universities within the nation's over-all research system? Research opportunities are increasing in size and complexity, often requiring large scale organizational settings for their performance. In addition, advances in fundamental knowledge are increasingly relevant to technology development.

What types of research will best be conducted by university-based scientists and engineers? What should be the roles of other research organizations, such as industrial, non-profit, and governmental laboratories? Should universities re-focus or narrow their priorities in research and education? Does the country need new types of research institutions to address scientific and technological opportunities and needs?

What is the appropriate role for universities in addressing national and regional priorities? Concern for improving the nation's international competitiveness has generated expectations that universities, in partnership with industry, will provide scientific and technological breakthroughs in key commercial areas. State and local officials increasingly urge their public universities to contribute to regional development through applied research and cooperation with resident industries.

What should be the relationship between national research-support policies and national, regional, state, and local economic-development policies? To what degree should universities respond to public expectations for them to address specific national or regional social, political, and economic priorities? How can universities maintain their independence while increasing their involvement in extramural research activities?

What should be the research and education role of U.S. universities within an increasingly international research environment? As more nations develop research capacity, U.S. research scientists and engineers will adjust their research priorities and programs to reflect dynamic worldwide changes in scientific fields. Furthermore, international cooperation is intensifying in many scientific and engineering fields. The growing research capabilities of other nations provide new opportunities for collaboration.

What is the appropriate balance between the global flow of scientific information and collaboration in research to advance scientific fields, on the one hand, and national policies to capture the economic and military benefits of scientific discoveries, on the other? With a shifting balance of international economic and scientific strength, should the U.S. target research areas with strategic importance or comparative advantage, and import from abroad the frontier scientific or technological knowledge developed within remaining fields?

Is there an optimal size, scope, and diversity of the U.S. academic research enterprise? Pressures of restricted funding, increased institutional competition, and steady-state enrollments generate concerns for maintaining excellence within the current academic research enterprise.

Should there be increased differentiation in research and education roles among institutions of higher education? How should the demands for maintaining research excellence be balanced with the exigencies for broadening participation? What are the trade-offs between concentrating research funding among the few, who have demonstrated success and quality, and allocating it more broadly among institutions in geographic regions?

Organization and Management of Universities

What are the implications of a changing research environment for university management and governance? As the environment outside the university is shifting and the nature of science and engineering research within the university is changing, university leadership will be increasingly challenged in its endeavor to maintain the pre-eminence of the enterprise.

Do the administrative, management, and governance structures of universities need to be modified to meet this challenge? What are pertinent models for future university administrative, management, and governance structures? As research on the scientific frontier evolves, crossing and extending disciplinary boundaries, how will the traditional departments adapt and how will their relationships with research centers, institutes, and other collaborative forms of organization develop? Should universities become involved in independent research efforts that require non-instructional personnel and depend on large-scale, sophisticated equipment and facilities? If so, how should such efforts be managed?

Does the current balance between scientific research and education require re-examination? The unique feature of the U.S. academic enterprise has long been its commitment to training scholars in research environments. There are increasing indications, however, that the link between teaching and research may be eroding, particularly as competition for research and development funds increases and research success becomes more closely associated with economic development. Other factors also challenge the teaching-research relationship, including declining graduate education support, changing faculty reward systems, and increasing use of "soft" money for faculty salaries.

How will the changing research climate affect teaching responsibilities and the quality of intellectual life at institutions of higher education? What are the implications of the evolving research environment at the classroom and laboratory level? Can the link between academic research and teaching be strengthened without compromising the missions of either? What core values and programs should not be neglected? What is the optimal balance between the sciences and engineering and the arts and humanities?

How should universities respond to the evolution of academic disciplines? During the past decade, many universities have begun to establish capacity for multi-disciplinary or inter-disciplinary research to meet emerging societal needs. In addition, new disciplines are emerging from older fields of inquiry. This development, however, has not been without problems--most notably, the faculty reward system that favors single-disciplinary research in established fields.

How can the universities restructure themselves to meet emerging societal expectations? With the current faculty reward system, can they devise methods to reward those who perform and publish in the multi-disciplinary arena? How can inter-disciplinary collaboration be enhanced? Will enhancing inter-disciplinary collaboration compromise freedom of inquiry?

Conduct of Research and Transfer of Knowledge

How will the scholarly agenda be set in the 1990s and by whom? The academic research agenda is guided by an increasingly complex array of influences. At a minimum, these include the precepts of the field of inquiry, the emergence of new technologies, and the social, political, and economic priorities of the country.

How can the research agenda be managed to preserve a balance between internal academic priorities and research opportunities, and external influences and needs? What is the best method for establishing priorities to allocate resources among disciplines, programs, and projects? How can appropriate output measures of academic research be developed to evaluate research productivity and efficiency? How can academic scientists and engineers participate in setting future research funding priorities?

As the research agenda evolves, how will it affect the role of investigators? Pressures for addressing political and socioeconomic priorities and for participating in larger scale research projects will increase. Future priorities among the modes of research--single investigator, small groups, multi-disciplinary centers--will be subject to intense debate.

How is the distribution of research support among these modes best determined and who should determine it? How can investigators respond best to changes in the way research problems are selected? To changes in the processes for carrying out research? To changes in the interactions among departments and disciplines, and with entities outside academia? To new and untried approaches to obtaining research support?

What are the ethical questions investigators should confront in the changing research environment? How should academic investigators with entrepreneurial activities balance those activities with their professorial and public service roles?

How should investigators respond to increased public concerns regarding toxic substances, release of genetically engineered material, and experiments involving animals or human subjects?

With steady-states in enrollments, faculty positions, and production of new doctoral researchers, can productivity per investigator be increased with more sophisticated electronic information networks, scientific instrumentation, and new forms of research organization?

Scientific and Technological Education

How can an adequate supply of students be attracted to careers in science and technology to meet the nation's personnel needs during the next century? Demographic declines in the college-age population and inadequate pre-college preparation in mathematics and sciences raise concerns not only among educators, but also among industrial leaders and others who depend on the availability of technical human resources.

Is the decline in numbers of students pursuing science and technological careers a problem systemic to the entire U.S. education system? If so, how can the kindergarten through high school system better motivate and educate creative and gifted young people? How can the educational system interest sufficient numbers of U.S. students in scientific and engineering careers to meet the future needs of industry, government, and universities?

How can an adequate supply of qualified scholars be attracted to academic research? Although the decline in the number of U.S. baccalaureate students who pursue advanced degrees appears to have turned around, the distribution of these students among the sciences may not provide sufficient numbers of qualified academic instructors and investigators in all disciplines.

What should be done to induce young talent, especially among women and minorities, to pursue careers in academic research? What incentives are necessary to encourage U.S. citizens to pursue careers in academic science and engineering?

How can the scientific literacy of U.S. citizens be advanced? Ultimately, national research priorities are greatly influenced by the concerns and perceived needs of citizens and their elected officials. The quality of those decisions depends, in large part, on an awareness of current scientific knowledge, an understanding of scientific methods, and an appreciation for the fiscal and organizational requirements of research.

What improvements can be made to the nation's education system to increase scientific and technical literacy among the citizenry? Are other methods available to increase effective and informed judgments on new scientific and engineering opportunities and their public policy implications? What role should educated lay audiences play in academic research? How can tensions be abated between those who perform research and those who influence public policy and, hence, research funding?

Funding Academic Research

How can sufficient resources for academic research be assured? The federal budget deficit and national reluctance to raise taxes indicate that all institutions that rely on substantial infusions of federal moneys for program funding will face increased difficulties in the 1990s. Coupled with the rising costs of research, the situation looms particularly arduous for research universities.

Should growth in funding for academic research be proportionate to growth in the nation's economy? How long can the past decade's high growth rate in academic research funds be sustained? What are the proper funding roles and responsibilities for the various sponsors of academic research--federal, state, and local agencies, industry, philanthropy, and the universities themselves? With multi-sponsor funding, how can meeting all enterprise needs--salaries, equipment, and facilities--be ensured? What is the appropriate balance between federal and non-federal funding levels. What is the appropriate balance between direct funding mechanisms and indirect mechanisms such as tax policy?

How should resources be allocated among competing national research objectives? New opportunities and demands for academic research occur with increasing frequency and intensity. This circumstance, coupled with growth of the number of institutions with basic research capacity, will pose difficult problems for allocating the finite research dollars that are available.

What methods should be used for setting funding priorities in research? How should the nation allocate resources between continuing investment in traditional research programs and underwriting new scientific approaches and combinations?

CONCLUSION

At the dawn of the 1990s, the United States is confronted with an academic research enterprise that shows the strains of rapid, dynamic growth and the consequences of its own success. In the last four decades, the nation has produced an academic research capability that is vastly larger and more decentralized than could have been foreseen by the most visionary policy-makers at the end of World War Two. The extraordinary success of the enterprise invites high ambitions for U.S. universities and colleges during the next decade. Powerful forces--within and without the university community--are generating pressures to further expand the role of academic research and broaden the institutional and geographic research base.

By pressing for an expansion of frontier research, as well as greater geographic diversity, the nation now faces decisions of how, to whom, to what extent, and for what purposes to allot limited resources. Sustaining the quality of current research institutions and programs will require increased financial and human resources, as well as organizational innovation. Policy-makers in government, industry, and universities will be forced to find an optimal balance among these competing demands and make pivotal investment and human-resource decisions that will profoundly influence the character and role of universities during the next century.

Maintaining the pre-eminence of the academic research enterprise will necessitate re-considering the major premises upon which it was established. Each university and college faces a range of choices, from accepting the challenge of an expanded mission to attempting to maintain its traditional role. For the enterprise as a whole, new strategies for its continued vitality must be considered--strategies far different from those employed by the research community, university administrators, and research sponsors in previous decades. Developing these strategies will test the nation's ingenuity and resourcefulness. The complexity of the issues, and the relationships among them, will require a comprehensive process and must involve all who hold a stake in the future of academic research.

NOTES

1. For a full discussion of the historical development of the academic research activities in other industrialized nations, see the symposium volume, *The University Research Enterprise within The Industrialized Nations: Comparative Historical Perspectives*, Government-University-Industry Research Roundtable, November, 1989.

2. Discussion of the research enterprise from 1890 to 1940 is derived from Roger L. Geiger, *To Advance Knowledge: The Growth of American Research Universities: 1900-1940,* New York: Oxford University Press, 1986.

3. See: Bush, Vannevar, *Science - The Endless Frontier: A Report to The President on A Program for Postwar Scientific Research*, July, 1945 (reprint: Washington: National Science Foundation, 1980); U.S. Congress, Senate Committee on Military Affairs, Subcommittee on War Mobilization, *Hearings on Science Legislation*, 1945 (Gilgore Report); and Steelman, John R., *Science and Public Policy: A Program for the Nation*, Washington, D.C.: U.S. Government Printing Office, 1947.

4. President's Science Advisory Committee, *Scientific Progress, The Universities, and The Federal Government*, 1960, pg.10-11. (the Seaborg Report).

5. National Academy of Sciences, *Federal Support of Basic Research in Institutions of Higher Learning*, 1964, pg. 92.

6. Research personnel (full-time equivalent) include those scientists and engineers (within the physical sciences, engineering, environmental sciences, life and health sciences, mathematics and computer sciences, and social and behavioral sciences) conducting funded (separately budgeted) academic R&D, estimated by the following: the fraction of faculty time spent in those research activities, non-faculty scientists and engineers employed to conduct research in campus facilities (except FFRDCs), post-doctoral researchers working in academic institutions, and graduate students paid as research assistants. As used in this report, doctoral universities are institutions that awarded an average of at least 10 Ph.D.s per year in the natural sciences and engineering between 1966 and 1986. There are 185 such institutions; 116 are public universities and 69 are private.

7. See Smith, Bruce L.R. and Joseph J. Karlesky, *The State of Academic Science*, New York: Change Magazine Press, 1977; Carnegie Corporation of New York et.al., *Research Universities and the National Interest: A Report from Fifteen University Presidents*, New York: Carnegie Corporation of New York, 1977; Sloan Commission on Government and Higher Education, "Federal Support for Academic Research," *A Program for Renewed Partnership*, New York: Sloan Foundation, 1980.

8. For public universities, such funds are in part derived from state and local government sources. Reported university-generated internal funds for research and development include institutional funds for separately budgeted research and development, cost-sharing, and under-recovery of indirect costs. They are derived from (1) general purpose state or local government appropriations, (2) general purpose grants from industry, foundations, or other outside sources, (3) tuition and fees, and (4) endowment income. See National Science Foundation, *Academic Science and Engineering R&D Funds,* 1987.

9. In 1986, unrecovered indirect R&D costs for public universities, as a percent of total R&D costs, was 10.7 percent--compared with 5.2 percent for private universities. Source: National Science Foundation, Division of Policy Research and Analysis.

10. See: *State Technology Programs in the United States: 1988*, Minnesota Department of Trade and Economic Development, Office of Science and Technology, 1988.

11. Compared to 4.7 percent annual growth for the top 20 universities. Source: National Science Foundation, Division of Policy Research and Analysis.

12. National Science Foundation, *Foreign Citizens in U.S. Science and Engineering: History, Status, and Outlook*, Washington, 1986.

13. In 1985, at the major private research universities, non-faculty appointments averaged 22 percent of doctoral (non-postdoctoral) personnel in the sciences and engineering, reaching 38 percent in physics and astronomy, 35 percent in computer science, and 40 percent in environmental sciences. For major public research universities, in 1985 non-faculty averaged 12 percent of employment of doctoral scientists and engineers. Source: Survey of Doctoral Recipients, National Research Council, Office of Scientific and Engineering Personnel.

14. National Science Foundation, SRS, special tabulations.

15. Assuming mid-level projections of the 18-to-22 year old cohort and current enrollment rates. U.S. Department of Commerce, Bureau of the Census, Current Population Reports, Series P-25, No. 952, "Projections of the Population of the United States by Age, Sex, and Race: 1983-2080."

16. To maintain the 1985 volume, the participation rate would have to increase to about 65 per thousand 22-year olds. For the last 15 years, participation rates have fluctuated between 40 and 50 per thousand. During the next decade, to maintain current levels of baccalaureate degrees in the sciences and engineering, a significant increase in the rate at which 22-year olds attain science and engineering degrees would be required. See: *Nurturing Science and Engineering Talent: A Discussion Paper*, Government-University-Industry Research Roundtable, July 1987.

17. See: *Nurturing Science and Engineering Talent: A Discussion Paper*, Government-University-Industry Research Roundtable, July 1987, pgs 7-12.

18. National Science Foundation, *Academic Research Equipment in Selected Science/Engineering Fields: 1982-1983 to 1985-1986*, 1988.

19. National Science Foundation, *Scientific and Engineering Research Facilities at Universities and Colleges*, September 1988.

20. Between 1980 and 1988, average compensation for academic research personnel (faculty and non-faculty) has increased by nearly 25 percent, accounting for inflation. Source: National Foundation, Division of Policy Research and Analysis.

SCIENCE AND TECHNOLOGY IN THE ACADEMIC ENTERPRISE

PART TWO:

OVERVIEW OF THE ACADEMIC RESEARCH ENTERPRISE

INTRODUCTION

During the past three decades, U.S. universities and colleges have assumed a major role in the nation's over-all research system. The academic research enterprise has grown dramatically--both in number of academic research personnel and in financial resources allocated to academic research. During the 1950s and 1960s, the growth of the enterprise was generally uniform in all its aspects: financial support, employment of academic personnel, university enrollments, and production of new scientists and engineers. During the past two decades, however, these trends have diverged, presenting policy-makers with a unique set of challenges.

Part Two of this discussion paper provides quantitative descriptions of the dynamic long-term trends which now affect the academic research enterprise. The Working Group hopes that this information will provide a necessary historical perspective to many of the current challenges facing the enterprise and add additional insights into many of the underlying influences which now shape its future. The quantitative information presented in this discussion paper primarily describes inputs to the academic research enterprise, such as financial and human resources. While some output measures have been developed--using publication and citation rates, patents, or departmental rankings--they require further methodological refinement before they can be meaningfully incorporated into analyses of academic research. Reliable data on long-term trends in academic research quality, productivity, or efficiency do not exist.

The charts in Part Two are derived from a database maintained by the National Science Foundation. Most of the data were produced from periodic national surveys of academic institutions conducted by the National Science Foundation and the U.S. Department of Education. In some instances, estimates have been incorporated within the database; this has been necessary for two reasons: first, not all of the survey instruments have consistently requested the same information in the same format; second, survey frequencies have changed, creating gaps in information for specific years. Additional information on the enterprise has been collected for specific years by federal agencies, philanthropic foundations, study commissions, professional associations, as well as individual investigators; the data from many of these studies have been used to supplement the survey data and to develop estimates where necessary.

The graphic information included here covers a three-decade time span, from 1958 through 1988. The data have been standardized to provide comparability among all the graphs; all financial data are expressed in 1988-constant dollars. It should be noted that descriptions of academic institutions are based on aggregated data for the entire enterprise or large sectors of it. Inferences for individual academic institutions should not be drawn from these data, as each university and college varies for all the characteristics described here.

During 1990, the Working Group will hold a series of conferences for university, congressional, federal, state, and industry officials, as well as academic scientists and engineers, to explore options and alternative scenarios for sustaining the quality of academic research during the 1990s and into the next century. The material in Part Two will be used as an information resource base for those conferences.

Part Two is divided into the following sections:

- Summary of major trends affecting the academic research enterprise.
- National research and development expenditures.
- National expenditures for academic research and development.
- Total academic expenditures and revenues.
- Academic personnel.
- Higher education enrollments.
- Science and engineering degrees.

SUMMARY OF MAJOR TRENDS

National R&D: Character

Total U.S. R&D expenditures in 1988 were more than $125 billion. Accounting for inflation, they have increased by about 400 percent since 1953. Basic research has increased sharply, from less than $3 billion (1988 dollars) in 1953 to more than $18 billion in 1989; as a result, its share of total R&D has risen from less than 10 percent to about 15 percent during the same period. Applied research has fluctuated between 20 percent and 25 percent; development has accounted for 65 percent to 70 percent of total R&D.

Figure 2-1: U.S. R&D Expenditures by Type of Research and Development

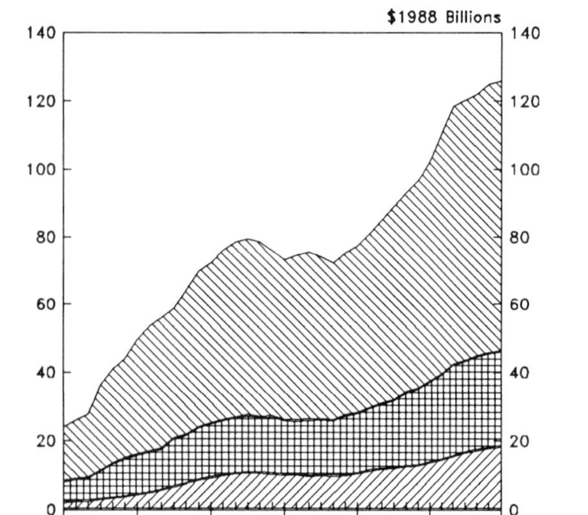

Figure 2-2: Distribution of R&D Expenditures by Type of Research and Development

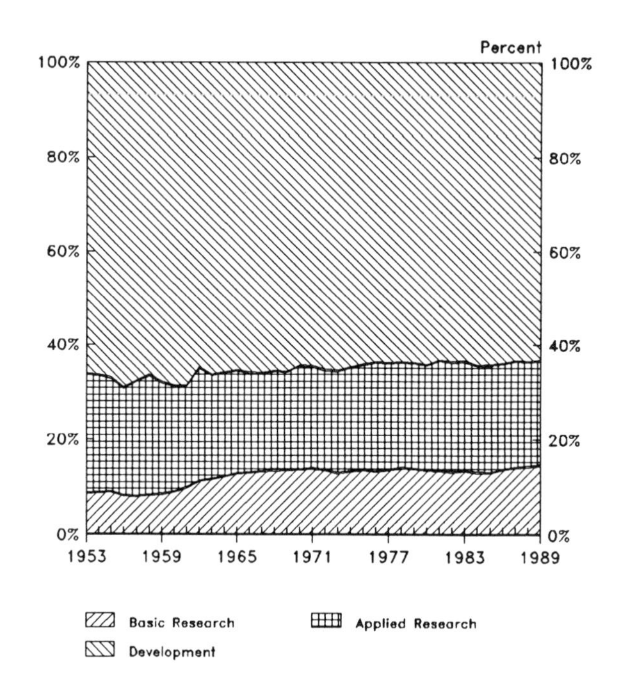

NOTE: Data series within the figures are not overlapped; top line represents total. Financial data are expressed in 1988 constant dollars to reflect real long-term growth trends.

DEFINITION OF TERMS: R&D expenditures include current-fund expenditures for all research and development activities that are separately budgeted and accounted for. *Basic Research* is the systematic study where the primary aim of the investigator is directed toward fuller knowledge or understanding of the subject under study, rather than a practical or commercial application thereof. *Applied Research* is the systematic study where the primary aim of the investigator is directed toward gaining knowledge or understanding necessary for determining the means by which a recognized and specific need or commercial objective may be met. *Development* is the systematic use of the knowledge or understanding gained from research, directed toward the production of useful materials, devices, systems, or methods, including design and development of prototypes and processes.

SOURCE: National Science Foundation, Division of Policy Research and Analysis. Database: CASPAR. Some of the data within this database are estimates, incorporated where there are discontinuities within data series or gaps in data collection. Primary data sources: National Science Foundation, Division of Science Resource Studies, Survey of Federal Funds for Research and Development; Survey of Federal Support to Universities, Colleges, and Non-profit Organizations, Survey of Scientific and Engineering Expenditures at Universities and Colleges; Survey of Industrial Research and Development.

Academic R&D: Share of Total U.S. R&D

During the 1960s, academic institutions assumed a more prominent role within the nation's over-all R&D system. Their share of U.S. basic research expenditures increased from 25 percent in 1953 to half by the early 1970s, where it has remained; their share of all basic and applied research went from 15 percent to 25 percent, that of total research and development from 5 percent to 10 percent.

Figure 2-3: Academic Share of U.S. R&D Expenditures

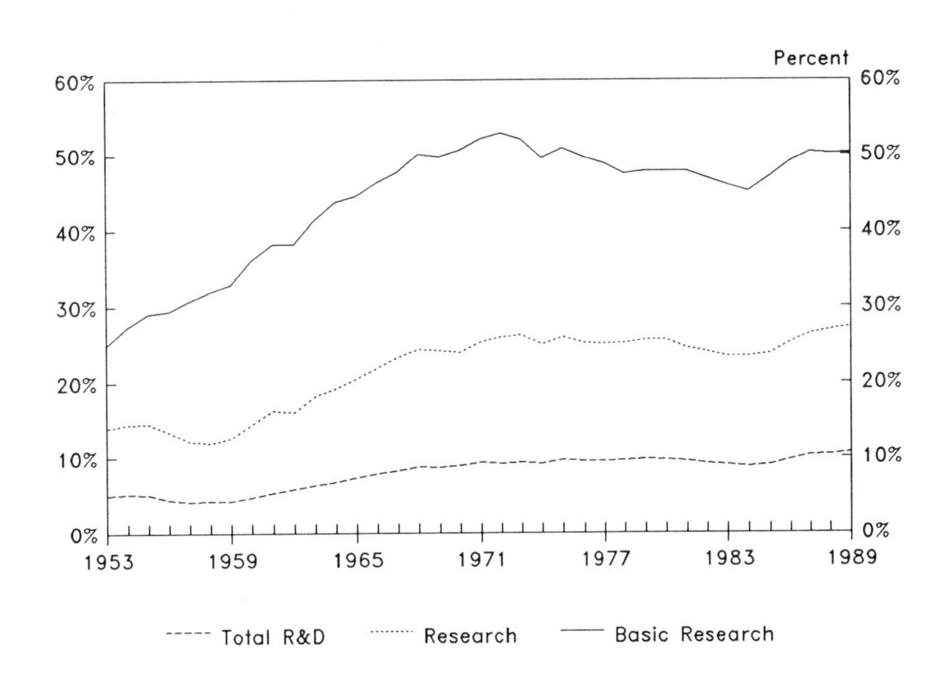

DEFINITION OF TERMS: Academic R&D Expenditures include current fund expenditures within higher education institutions for all research and development activities that are separately budgeted and accounted for. This includes both sponsored research activities (sponsored by federal and non-federal agencies and organizations) and university research separately budgeted under an internal application of institutional funds; and excludes training grants, public service grants, demonstration projects, and departmental research expenditures that are not separately budgeted. *Total R&D* includes all non-capital national expenditures for the conduct of basic research, applied research, and development. *Research* includes all non-capital national expenditures for basic and applied research. *Basic Research* includes all non-capital national expenditures for the conduct of basic research. Basic research is the systematic study where the primary aim of the investigator is directed toward fuller knowledge or understanding of the subject under study, rather than a practical or commercial application thereof. Applied research is the systematic study where the primary aim of the investigator is directed toward gaining knowledge or understanding necessary for determining the means by which a recognized and specific need or commercial objective may be met. Development is the systematic use of the knowledge or understanding gained from research, directed toward the production of useful materials, devices, systems, or methods, including design and development of prototypes and processes.

SOURCE: National Science Foundation, Division of Policy Research and Analysis. Database: CASPAR. Some of the data within this database are estimates, incorporated where there are discontinuities within data series or gaps in data collection. Primary data sources: National Science Foundation, Division of Science Resource Studies, Survey of Federal Funds for Research and Development; Survey of Federal Support to Universities, Colleges, and Non-profit Organizations; Survey of Scientific and Engineering Expenditures at Universities and Colleges; Survey of Industrial Research and Development.

Academic R&D: Share of U.S. GNP

Academic R&D as a percentage of the nation's gross national product rose sharply and continuously during the 1950s and 1960s, from 0.07 percent in 1953 to 0.25 percent by 1968; after falling to 0.21 percent in the 1970s, it has reached a new high of 0.27 percent in the late 1980s. The federal funding share of academic R&D grew from 0.04 percent in 1953 to 0.17 percent by 1968; after declining during 1970s, it returned to 0.16 percent by 1988.

Figure 2-4: Total and Federal Academic R&D Funds as Percents of U.S. GNP

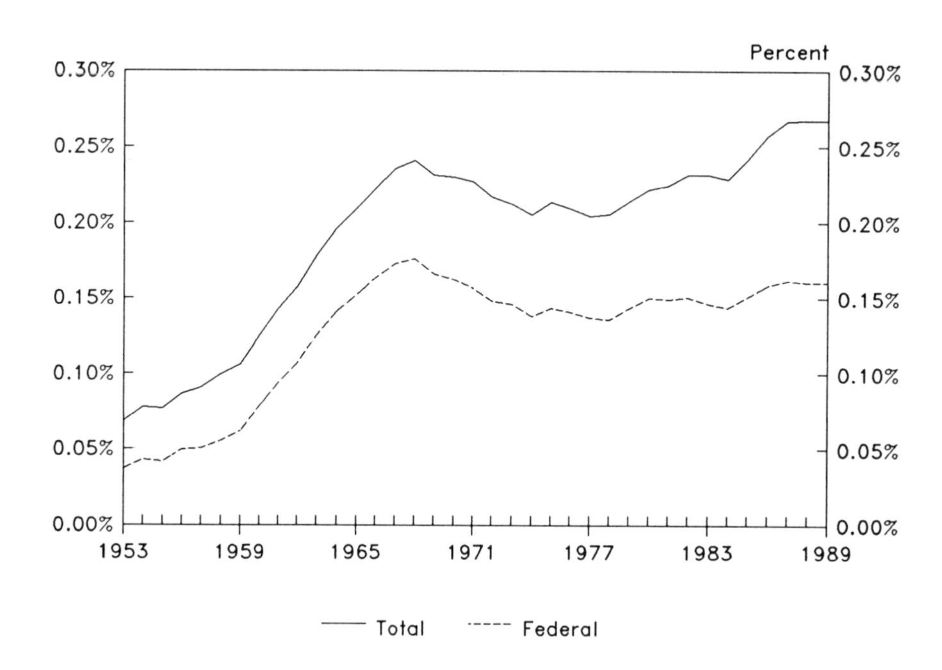

DEFINITION OF TERMS: *Total* academic R&D expenditures include current-fund expenditures within higher education institutions for all research and development activities that are separately budgeted and accounted for. This includes both sponsored research activities (sponsored by federal and non-federal agencies and organizations) and university research separately budgeted under an internal application of institutional funds; and excludes training grants, public service grants, demonstration projects, and departmental research expenditures that are not separately budgeted. *Federal* funds include grants and contracts to academic institutions for R&D (including direct and reimbursed indirect costs) by agencies of the federal government; excludes funds for FFRDCs. Gross national product is the estimated total market value of all goods and services produced annually in the United States.

SOURCE: National Science Foundation, Division of Policy Research and Analysis. Database: CASPAR. Some of the data within this database are estimates, incorporated where there are discontinuities within data series or gaps in data collection. Primary data sources: National Science Foundation, Division of Science Resource Studies, Survey of Scientific and Engineering Expenditures at Universities and Colleges; U.S. Department of Commerce, Bureau of Economic Analysis, Survey of Current Business and Commerce.

Doctoral Institution Growth Patterns: R&D Expenditures

An index of total and federal funding of R&D within doctoral institutions reveals a pattern of strong growth during the 1950s and 1960s, little or no growth in the 1970s, and strong increases in the 1980s.

Figure 2-5: Index of Doctoral Institution Total and Federal R&D Funds

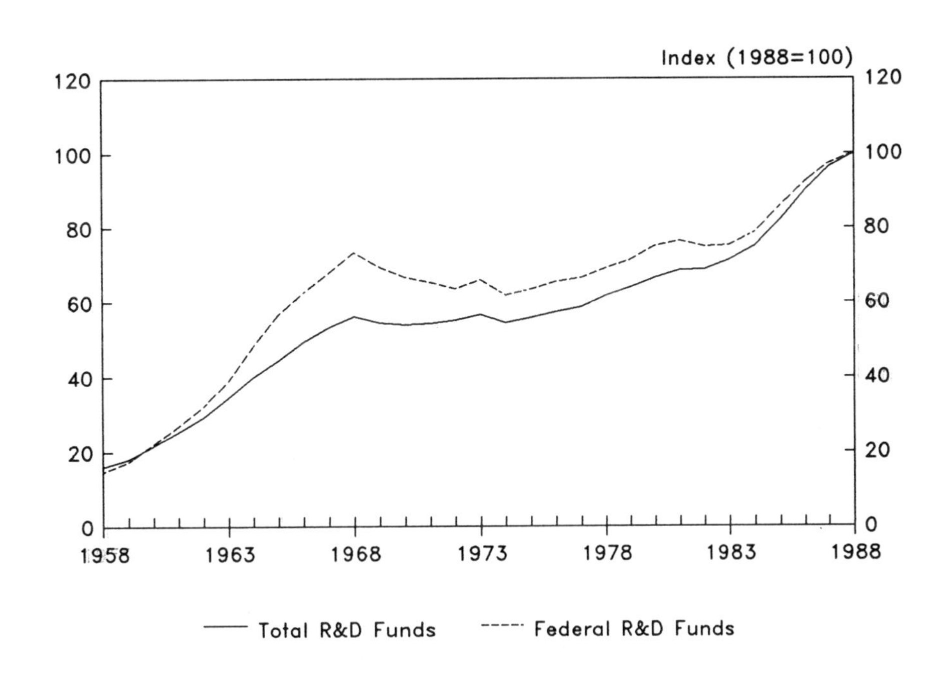

NOTE: Index based on financial data computed in 1988 constant dollars.

DEFINITION OF TERMS: *Total R&D Funds* include all current-fund expenditures within doctoral institutions for all research and development activities that are separately budgeted and accounted for. This includes both sponsored research activities (sponsored by federal and non-federal agencies and organizations) and university research separately budgeted under an internal application of institutional funds; and excludes training, public service, demonstration projects, and departmental research separately budgeted and FFRDCs. *Federal R&D Funds* include grants and contracts by agencies of the federal government for R&D (including direct and reimbursed indirect costs) made to doctoral institutions; excludes funds for FFRDCs. Doctoral institutions are institutions that have granted an average of 10 or more Ph.D. degrees per year in the natural sciences or engineering over the past two decades. They include 116 public and 69 private institutions.

SOURCE: National Science Foundation, Division of Policy Research and Analysis. Database: CASPAR. Some of the data within this data base are estimates, incorporated where there are discontinuities within data series or gaps in data collection. Primary data source: National Science Foundation, Division of Science Resource Studies, Survey of Scientific and Engineering Expenditures at Universities and Colleges.

Doctoral Institution Growth Patterns: Revenues and Expenditures

An index of total operating revenues and expenditures for doctoral institutions reveals strong, steady real growth during the past three decades.

Figure 2-6: Index of Doctoral Institution Operating Revenues and Expenditures

Index (1988=100)
120
100
80
60
40
20
0
1958
1963
1968
1973
1978
1983
1988
Revenues
Expenditures

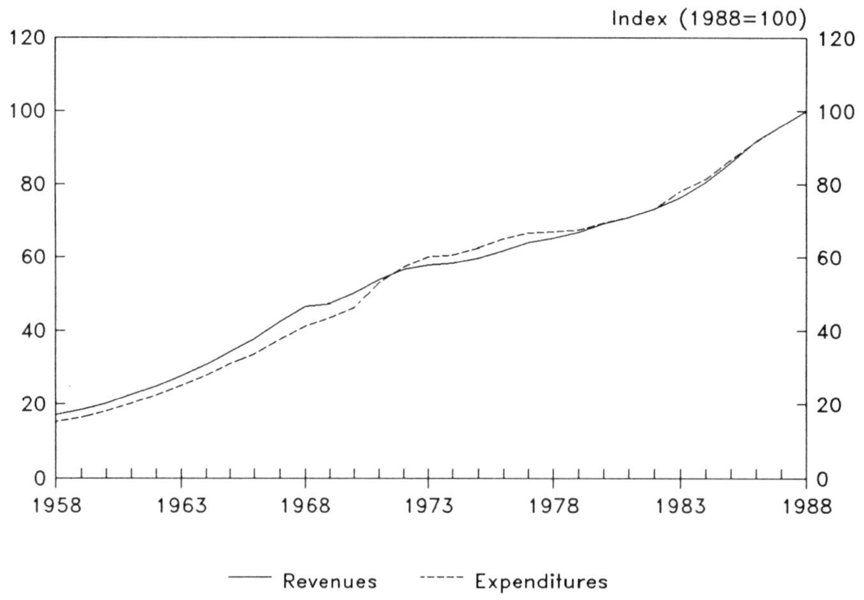

NOTE: Index based on financial data computed in 1988 constant dollars.

DEFINITION OF TERMS: *Revenues* consist of current-fund revenues from federal, state, and local appropriations; tuition income, government grants and contracts; private gifts, grants, and endowment income; sales and services of educational activities; and revenues from hospitals, auxiliary enterprise, and FFRDCs. Excluded are revenues for capital purposes and Pell Grants. Doctoral institutions include those institutions which have granted an average of 10 or more Ph.D. degrees per year in the natural sciences or engineering over the past two decades. They include 116 public and 69 private institutions. *Expenditures* consist of current-fund expenditures for instruction, research, public service, academic support, student services, institutional support, operation and maintenance of plant, scholarships and fellowships, and educational and mandatory transfers and expenditures for hospitals, auxiliary enterprises, and FFRDCs. Excludes expenditures from institutional plant fund accounts.

SOURCE: National Science Foundation, Division of Policy Research and Analysis. Database: CASPAR. Some of the data within this database are estimates, incorporated where there are discontinuities within data series or gaps in data collection. Primary data source: U.S. Department of Education, National Center for Education Statistics, Higher Education General Information Survey (HEGIS): Financial Statistics of Institutions of Higher Education.

Doctoral Institution Growth Patterns: Personnel

An index of doctoral institution personnel reveals strong growth for total faculty until 1970s, and no growth in the 1980s; for scientists and engineers, it shows uninterrupted growth for three decades. For research personnel it shows growth through the 1960s, a levelling off in the 1970s, and strong increases in 1980s.

Figure 2-7: Index of Doctoral Institution Employment of Total Faculty, FTE Scientists and Engineers, and FTE Investigators

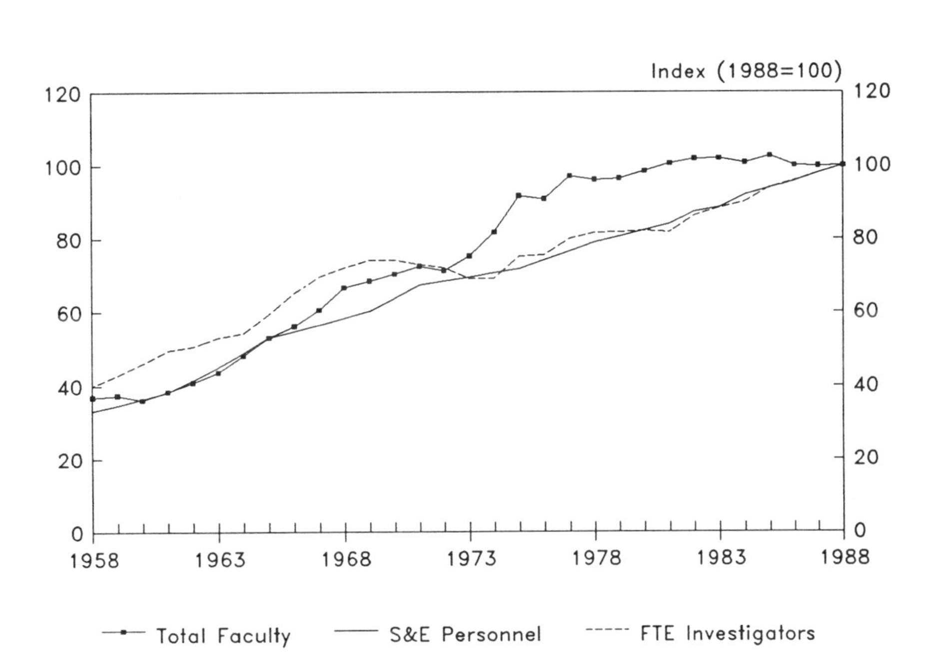

DEFINITION OF TERMS: *Total Faculty* include all instructional members of the instruction or research staff of doctoral institutions whose major regular assignment is instruction, including those with release time for research. *S&E Personnel* include all scientists and engineers including both faculty and non-faculty personnel and post-doctorates, employed by higher education institutions (plus a full-time equivalent for part-time employees), within the following broad fields: physical sciences, engineering, environmental sciences, life and health sciences, mathematics and computer sciences, and social and behavioral sciences. *FTE Investigators* (full-time equivalent) include those scientists and engineers conducting funded (separately budgeted) academic R&D; the full-time equivalent is an estimate, derived from the fraction of faculty time spent in those research activities, non-faculty scientists and engineers employed to conduct research in campus facilities (except FFRDCs), and post-doctoral researchers working in academic institutions.

SOURCE: National Science Foundation, Division of Policy Research and Analysis. Database: CASPAR. Some of the data within this database are estimates, incorporated where there are discontinuities within data series or gaps in data collection. Primary data sources: U.S. Department of Education, National Center for Education Statistics, Higher Education General Information Survey (HEGIS): Salaries, Tenure, and Fringe Benefits of Full-time Instructional Faculty; American Council on Education; National Science Foundation, Division of Science Resource Studies, Survey of Scientific and Engineering Personnel Employed at Universities and Colleges.

Doctoral Institution Growth Patterns: Enrollments

An index of total and graduate student enrollments reveals strong growth until the mid-1970s and little growth thereafter.

Figure 2-8: Index of Total and Graduate Enrollment in Doctoral Institutions

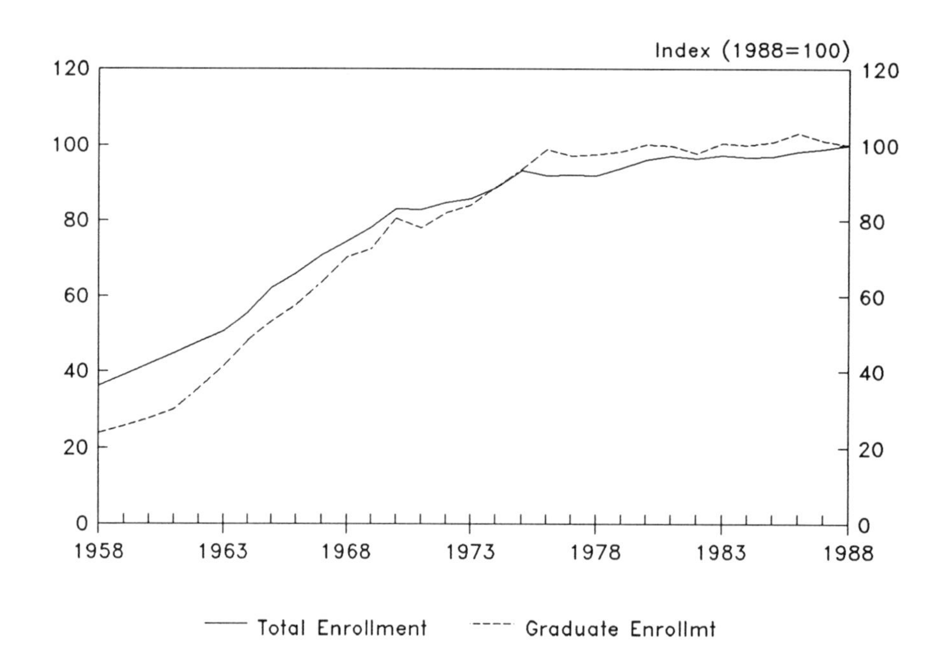

DEFINITION OF TERMS: *Total Enrollment* include all full-time students (plus a full-time equivalent of part-time students) as reported by doctoral institutions. *Graduate Enrollment* include all full-time students (plus a full-time equivalent of part-time students) who hold a bachelors degree, or equivalent, and are working toward an advanced degree including a first professional degree. Doctoral institutions are higher education institutions that have granted an average of 10 or more Ph.D. degrees per year in the natural sciences or engineering over the past two decades. They include 116 public and 69 private institutions.

SOURCE: National Science Foundation, Division of Policy Research and Analysis. Database: CASPAR. Some of the data within this data base are estimates, incorporated where there are discontinuities within data series or gaps in data collection. Primary data source: U.S. Department of Education, National Center for Education Statistics, Higher Education General Information Survey (HEGIS): Fall Enrollment in Institutions of Higher Education.

Doctoral Institution Growth Patterns: S&E Degrees

An index of bachelors degrees in the sciences and engineering, granted by doctoral institutions, reveals strong growth until the mid-1970s, then slowing growth during the 1980s. The growth in Ph.D. degrees was also steep during the 1960s, with decline during 1970s, and a return to early-1970s levels by 1988.

Figure 2-9: Index of Doctoral Institution Ph.D. and Bachelors Degrees Awarded in Science and Engineering

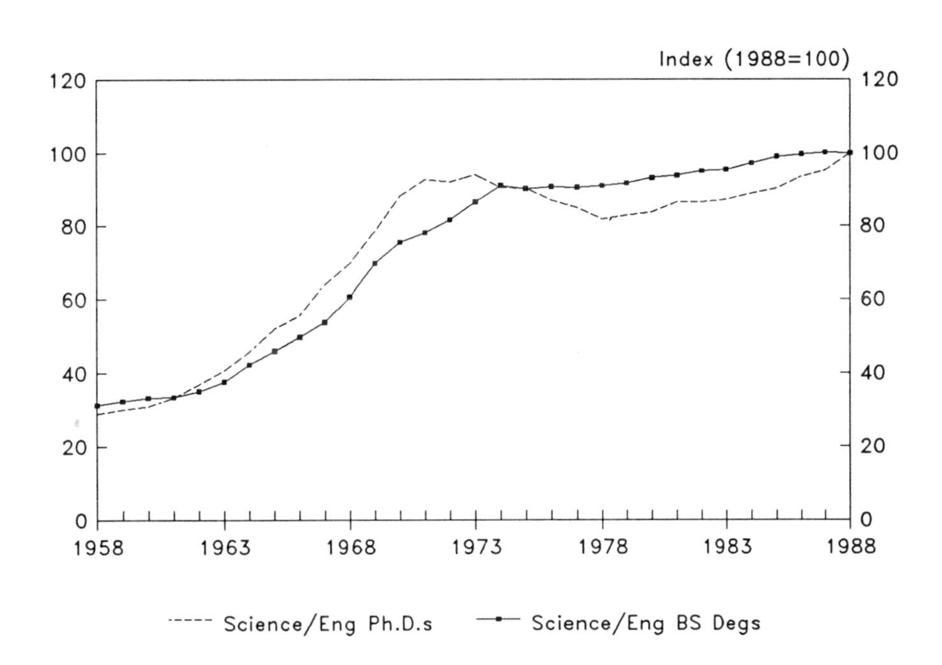

DEFINITION OF TERMS: *Science/Engineering Ph.D.s* and *Science/Enginerring B.S. Degrees* include those in life sciences, including agricultural, biological, medical, and other health sciences; physical sciences, including astronomy, chemistry, and physics; engineering, including aeronautical and astronautical, chemical, civil, electrical, and mechanical engineering; environmental sciences, including oceanography, atmospheric and earth sciences; mathematics and computer science, including all fields of mathematics and computer-related sciences; and social and other, including economics, political science, psychology, and sociology. Doctoral institutions are higher education institutions that have granted an average of 10 or more Ph.D. degrees per year in the natural sciences or engineering over the past two decades. They include 116 public and 69 private institutions.

SOURCE: National Science Foundation, Division of Policy Research and Analysis. Database: CASPAR. Some of the data within this database are estimates, incorporated where there are discontinuities within data series or gaps in data collection. Primary data source: U.S. Department of Education, National Center for Education Statistics, Higher Education General Information Survey (HEGIS): Degrees and Other Formal Awards Conferred.

Doctoral Institution Growth Patterns: Per-Person Expenditures

An index of average operating expenditures per faculty member, as well as education expenditures per student and per degree granted, reveals steady growth during the 1960s, no growth during the 1970s, then rapid growth through the 1980s.

Figure 2-10: Index of Doctoral Institution Per-Unit Expenditures

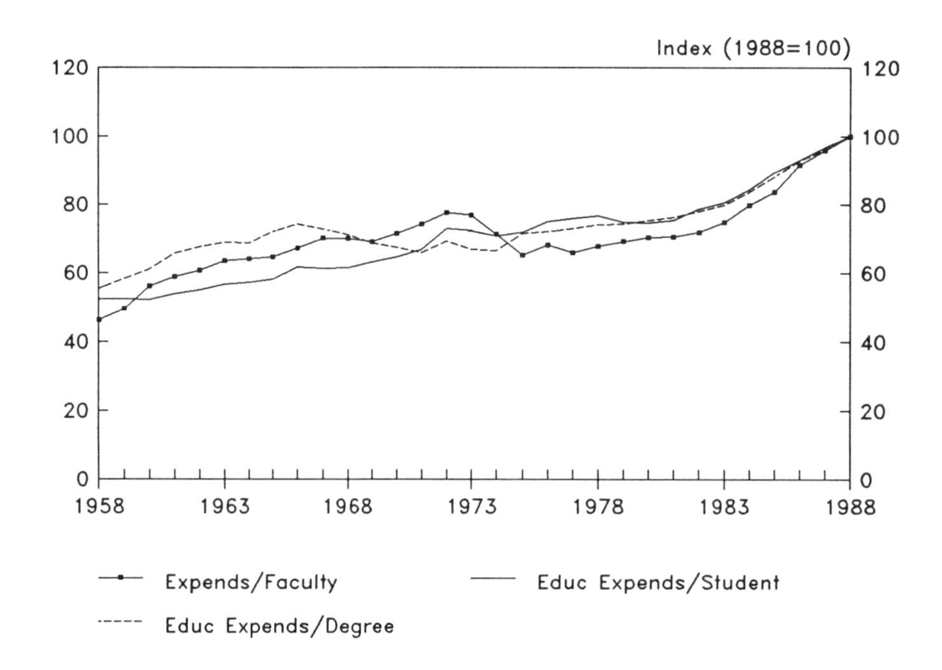

NOTE: Index based on financial data computed in 1988 constant dollars.

DEFINITION OF TERMS: *Expenditures* consist of current-fund expenditures for instruction, research, public service, academic support, student services, institutional support, operation and maintenance of plant, scholarships and fellowships, and educational and mandatory transfers and expenditures for hospitals, auxiliary enterprises, and FFRDCs. Excludes expenditures from institutional plant fund accounts and Pell Grants. *Educational Expenditures* include current-fund expenditures for instruction, academic support, student services, institutional support, operation and maintenance of plant, scholarships and fellowships, and educational and mandatory transfers. *Faculty* include all instructional members of the instruction or research staff whose major regular assignment is instruction, including those with release time for research. *Students* include all full-time students plus a full-time equivalent of part-time students as reported by doctoral institutions. *Degrees* include all degrees--undergraduate and graduate--in all academic disciplines. Doctoral institutions are institutions that have granted an average of 10 or more Ph.D. degrees per year in the natural sciences or engineering over the past two decades. They include 116 public and 69 private institutions.

SOURCE: National Science Foundation, Division of Policy Research and Analysis. Database: CASPAR. Some of the data within this database are estimates, incorporated where there are discontinuities within data series or gaps in data collection. Primary data sources: U.S. Department of Education, National Center for Education Statistics, Higher Education General Information Survey (HEGIS): Degrees and Other Formal Awards Conferred, Fall Enrollment in Institutions of Higher Education, Financial Statistics of Institutions of Higher Education; American Council on Education; National Association of State Universities and Land Grant Colleges.

NATIONAL R&D EXPENDITURES

National R&D: Performers

During thc 1960s, academic institutions increased their share of total national R&D expenditures, from 5 percent to 10 percent, where it has remained. By 1988, total U.S. R&D expenditures had risen to over $125 billion.

Figure 2-11: U.S. R&D Expenditures by Performer

$1988 Billions

Academic
Federal
Industry
Other

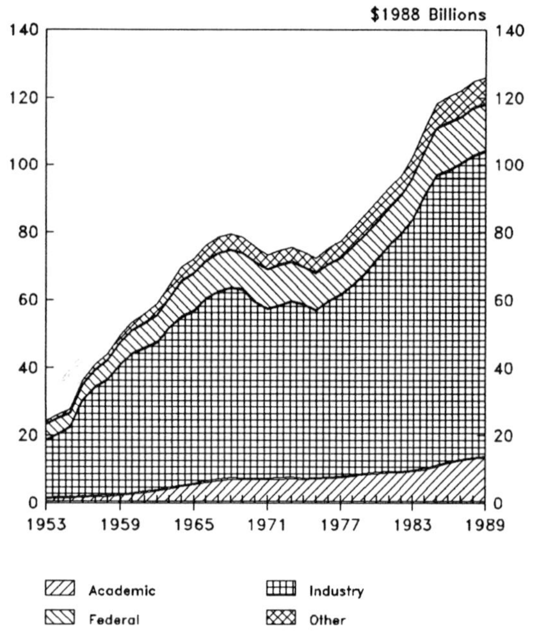

Figure 2-12: Distribution of U.S. R&D Expenditures by Performer

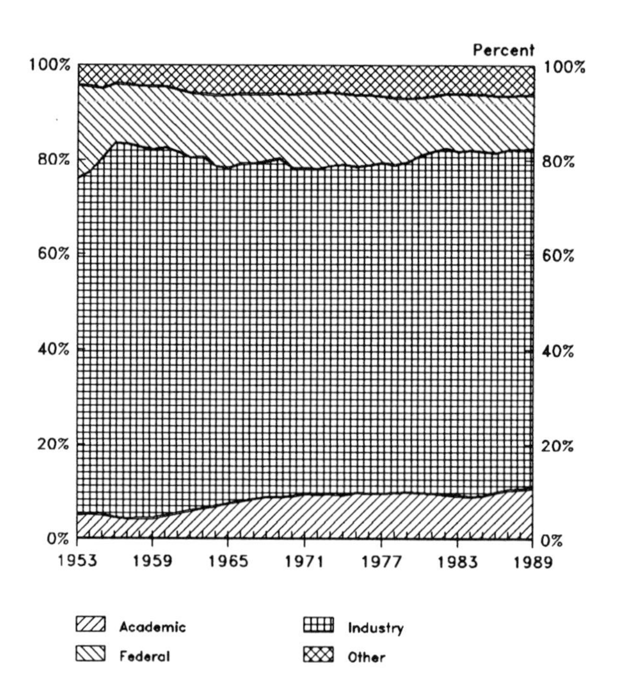

NOTE: Data series within the figures are not overlapped; top line represents total. Financial data are expressed in 1988 constant dollars to reflect real long-term growth trends.

DEFINITION OF TERMS: R&D expenditures include current-fund expenditures for all research and development activities that are separately budgeted and accounted for. *Academic* sector consists of public and private institutions of higher education including 185 doctoral, 1,224 comprehensive, and 1,388 2-year institutions; federally funded research and development centers (FFRDCs) administered by universities are reported under the Other category. *Industry* sector consists of both manufacturing and non-manufacturing companies; FFRDCs administered by industry are reported within this category. *Federal* sector consists of all agencies of the federal government. *Other* sector consists of public and private non-profit organizations that are involved in performing R&D, including FFRDCs administered by non-profit organizations.

SOURCE: National Science Foundation, Division of Policy Research and Analysis. Database: CASPAR. Some of the data within this database are estimates, incorporated where there are discontinuities within data series or gaps in data collection. Primary data sources: National Science Foundation, Division of Science Resource Studies, Survey of Federal Funds for Research and Development; Survey of Federal Support to Universities, Colleges, and Non-profit Organizations; Survey of Scientific and Engineering Expenditures at Universities and Colleges; Survey of Industrial Research and Development.

National R&D: Sources of Funding

Since the mid-1960s, the federal share of support for total national R&D expenditures has declined, from 65 percent in 1968 to less than half in 1988. During the same period, industry's share has grown, from 30 percent in 1968 to nearly 50 percent in 1988.

Figure 2-13: U.S. R&D Expenditures by Source of Funds

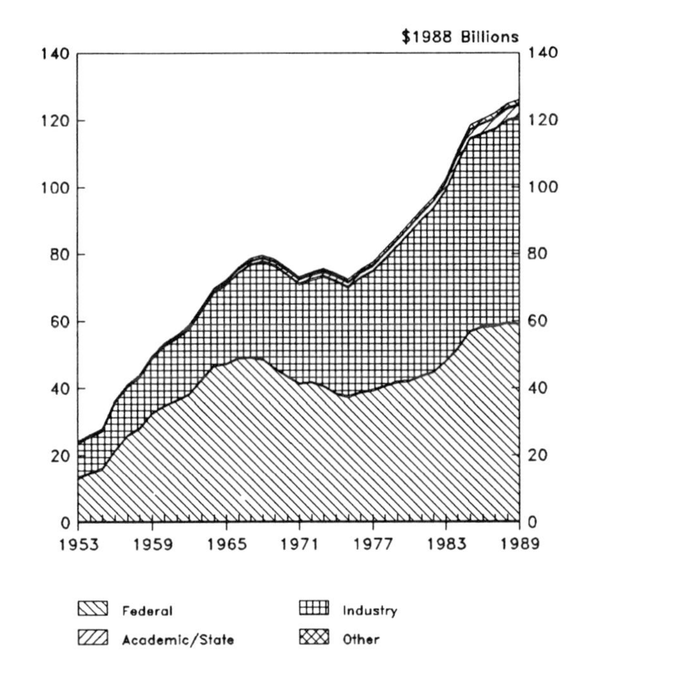

Figure 2-14: Distribution of U.S. R&D Expenditures by Source of Funds

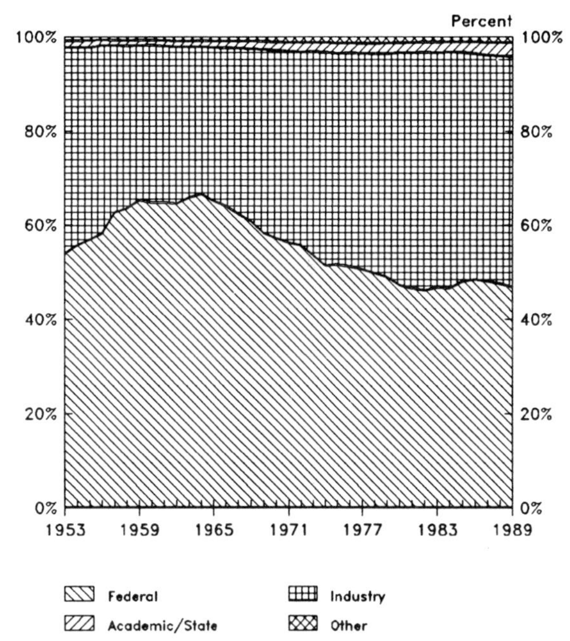

NOTE: Data series within the figures are not overlapped; top line represents total. Financial data are expressed in 1988 constant dollars to reflect real long-term growth trends.

DEFINITION OF TERMS: Research is the systematic study directed toward fuller knowledge or understanding (basic and applied) of the subject studied; development is systematic use of the knowledge or understanding gained from research, directed toward the production of useful materials, devices, systems, or methods, including design and development of prototypes and processes. R&D expenditures include current-fund expenditures for all research and development activities that are separately budgeted and accounted for. *Federal* sector consists of all agencies of the federal government. *Industry* sector consists of both manufacturing and non-manufacturing companies; industry funding of industrial research includes all funds (e.g. state and local) other than those received from the federal government. *Academic/State* funding of research and development includes general educational funds, from any source, that academic institutions have been free to allocate for separately budgeted research; and state and local government funds separately budgeted for academic R&D. *Other* sector consists of institutions that are primarily granting in nature, such as private philanthropic foundations and voluntary health agencies.

SOURCE: National Science Foundation, Division of Policy Research and Analysis. Database: CASPAR. Some of the data within this database are estimates, incorporated where there are discontinuities within data series or gaps in data collection. Primary data sources: National Science Foundation, Division of Science Resource Studies, Survey of Federal Funds for Research and Development; Survey of Federal Support to Universities, Colleges, and Non-profit Organizations; Survey of Scientific and Engineering Expenditures at Universities and Colleges; Survey of Industrial Research and Development.

National Research: Performers

For the past two decades, academic institutions have maintained a 25-percent share of total national research (basic and applied) expenditures. By 1988, total research expenditures had risen to $46 billion.

Figure 2-15: U.S. Research Expenditures by Performer

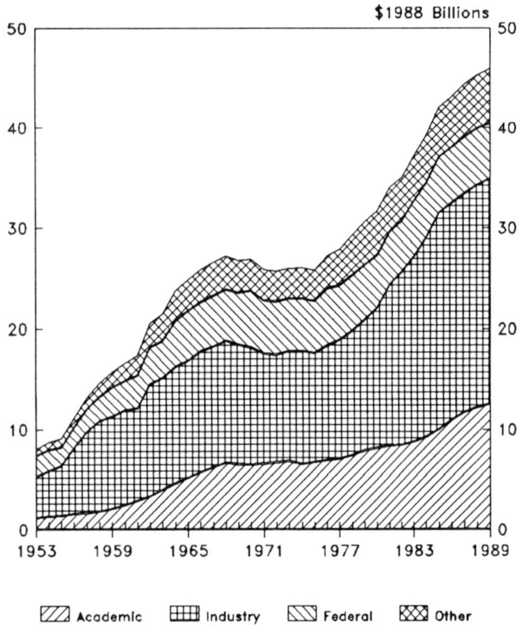

Figure 2-16: Distribution of U.S. Research Expenditures by Performer

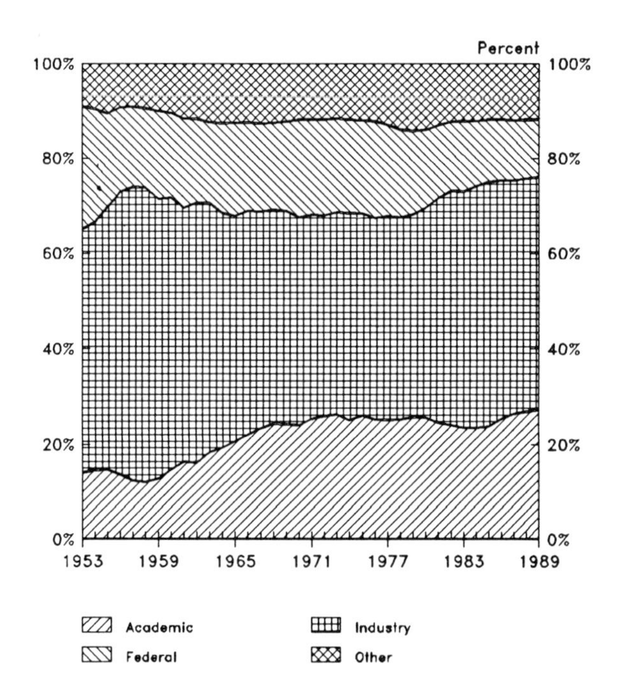

NOTE: Data series within the figures are not overlapped; top line represents total. Financial data are expressed in 1988 constant dollars to reflect real long-term growth trends.

DEFINITION OF TERMS: Research is a systematic study directed toward fuller knowledge or understanding (basic and applied) of the subject studied. Research expenditures include current-fund expenditures for all research activities that are separately budgeted and accounted for. *Academic* sector consists of public and private institutions of higher education, including 185 doctoral, 1,224 comprehensive, and 1,388 2-year institutions; (federally funded research and development centers (FFRDCs) administered by universities are reported under the Other category). *Industry* sector consists of both manufacturing and non-manufacturing companies; FFRDCs administered by industry are reported within this category. *Federal* sector consists of all agencies of the federal government. *Other* sector consists of public and private non-profit organizations that are involved in performing R&D, including FFRDCs administered by non-profit organizations.

SOURCE: National Science Foundation, Division of Policy Research and Analysis. Database: CASPAR. Some of the data within this database are estimates, incorporated where there are discontinuities within data series or gaps in data collection. Primary data sources: National Science Foundation, Division of Science Resource Studies, Survey of Federal Funds for Research and Development; Survey of Federal Support to Universities, Colleges, and Non-profit Organizations; Survey of Scientific and Engineering Expenditures at Universities and Colleges; Survey of Industrial Research and Development.

National Research: Sources of Funding

Since the mid-1960s, the federal share of support for total national research (basic and applied) expenditures has declined from 62 percent in 1968 to 52 percent in 1988, while industry's share has grown from 33 percent in 1968 to 38 percent in 1988.

Figure 2-17: U.S. Research Expenditures by Source of Funds

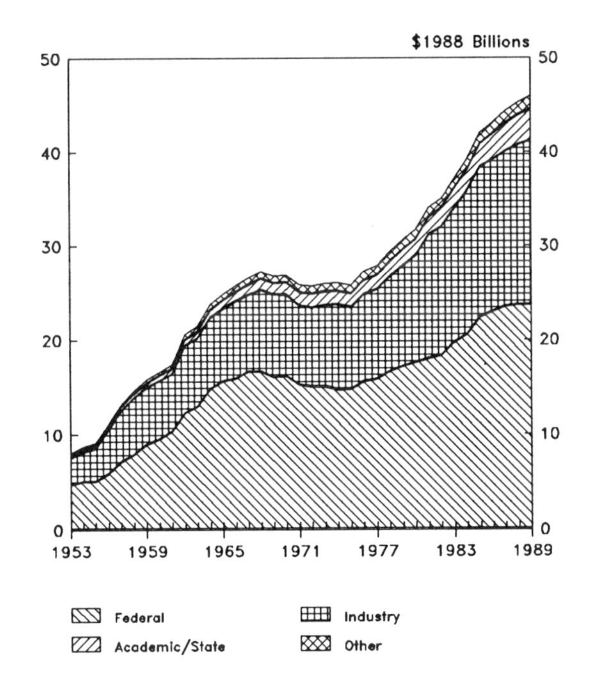

Figure 2-18: Distribution of U.S. Research Expenditures by Source of Funds

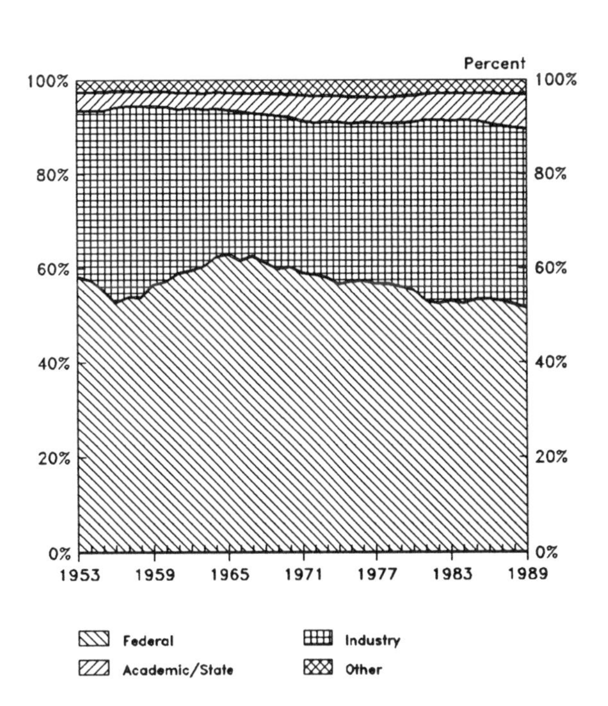

NOTE: Data series within the figures are not overlapped; top line represents total. Financial data are expressed in 1988 constant dollars to reflect real long-term growth trends.

DEFINITION OF TERMS: Research is the systematic study directed toward fuller knowledge or understanding (basic and applied) of the subject studied. Research expenditures include current-fund expenditures for all research activities that are separately budgeted and accounted for. *Federal* sector consists of all agencies of the federal government. *Industry* sector consists of both manufacturing and non-manufacturing companies; industry funding of industrial research includes all funds (e.g. state and local) other than those received from the federal government. *Academic/State* funding of research and development includes general educational funds, from any source, that academic institutions have been free to allocate for separately budgeted research; and state and local government funds separately budgeted for R&D. *Other* sector consists of institutions that are primarily granting in nature, such as private philanthropic foundations and voluntary health agencies.

SOURCE: National Science Foundation, Division of Policy Research and Analysis. Database: CASPAR. Some of the data within this database are estimates, incorporated where there are discontinuities within data series or gaps in data collection. Primary data sources: National Science Foundation, Division of Science Resource Studies, Survey of Federal Funds for Research and Development; Survey of Federal Support to Universities, Colleges, and Non-profit Organizations; Survey of Scientific and Engineering Expenditures at Universities and Colleges; Survey of Industrial Research and Development.

National Basic Research: Performers

Academic institutions have assumed a prominent role in the conduct of the nation's basic research, increasing their over-all share of basic research expenditures from 25 percent in 1953 to 50 percent by the early-1970s; after declining to 45-percent share in the late 1970s, academic institutions resumed 50-percent share by the late 1980s. By 1988, total U.S. basic research expenditures had risen to over $18 billion.

Figure 2-19: U.S. Basic Research Expenditures by Performer

$1988 Billions

0 5 10 15 20

1953 1959 1965 1971 1977 1983 1989

Academic Industry Federal Other

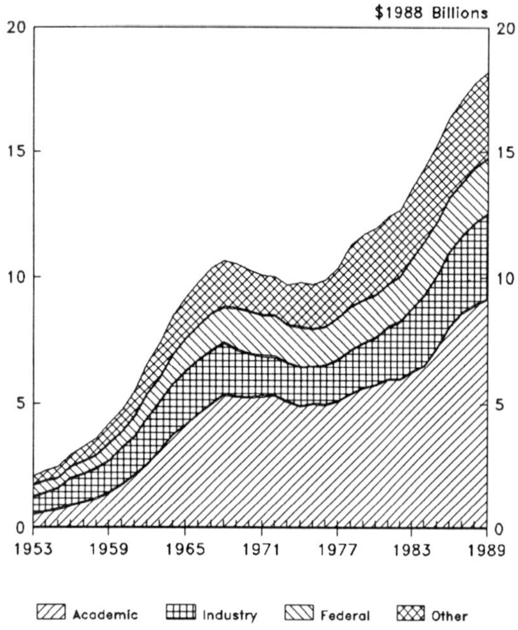

Figure 2-20: Distribution of U.S. Basic Research Expenditures by Performer

NOTE: Data series within the figures are not overlapped; top line represents total. Financial data are expressed in 1988 constant dollars to reflect real long-term growth trends.

DEFINITION OF TERMS: Basic research is a systematic study where the primary aim of the investigator is directed toward fuller knowledge or understanding of the subject under study, rather than a practical or commercial application thereof. Research expenditures include current-fund expenditures for all research activities that are separately budgeted and accounted for. *Academic* sector consists of public and private institutions of higher education, including 185 doctoral, 1,224 comprehensive, and 1,388 2-year institutions; federally funded research and development centers (FFRDCs) administered by universities are reported under the Other category. *Industry* sector consists of both manufacturing and non-manufacturing companies; FFRDCs administered by industry are reported in this category. *Federal* sector consists of all agencies of the federal government. *Other* sector consists of public and private non-profit organizations that are involved in performing R&D, including FFRDCs administered by non-profit organizations.

SOURCE: National Science Foundation, Division of Policy Research and Analysis. Database: CASPAR. Some of the data within this database are estimates, incorporated where there are discontinuities within data series or gaps in data collection. Primary data sources: National Science Foundation, Division of Science Resource Studies, Survey of Federal Funds for Research and Development; Survey of Federal Support to Universities, Colleges, and Non-profit Organizations; Survey of Scientific and Engineering Expenditures at Universities and Colleges; Survey of Industrial Research and Development.

National Basic Research: Sources of Funding

The federal share of support for basic research grew from 55 percent in 1958 to 70 percent in the 1960s and 1970s; declining to 64 percent by 1988. The industrial support share declined during the 1960s, from 33 percent in 1958 to 15 percent in 1978; rising to 20 percent by 1988. The share contributed together by academic institutions and state and local governments has increased from 2 percent in 1953 to over 12 percent in 1988.

Figure 2-21: U.S. Basic Research Expenditures by Source of Funds

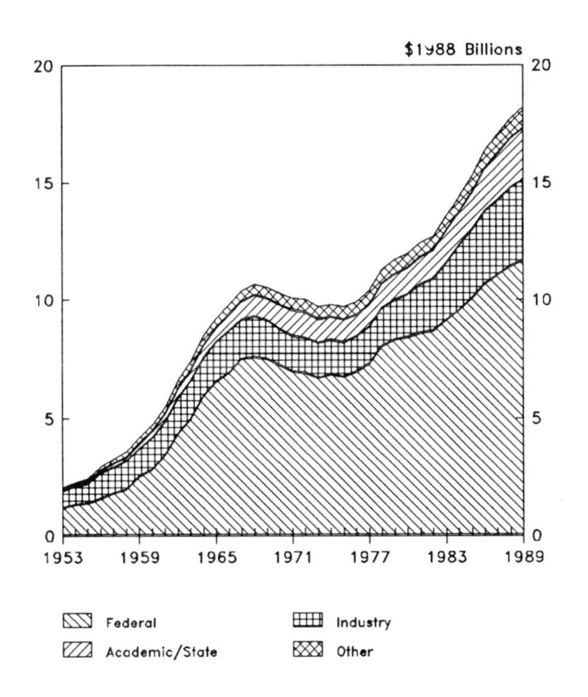

Figure 2-22: Distribution of U.S. Basic Research Expenditures by Source of Funds

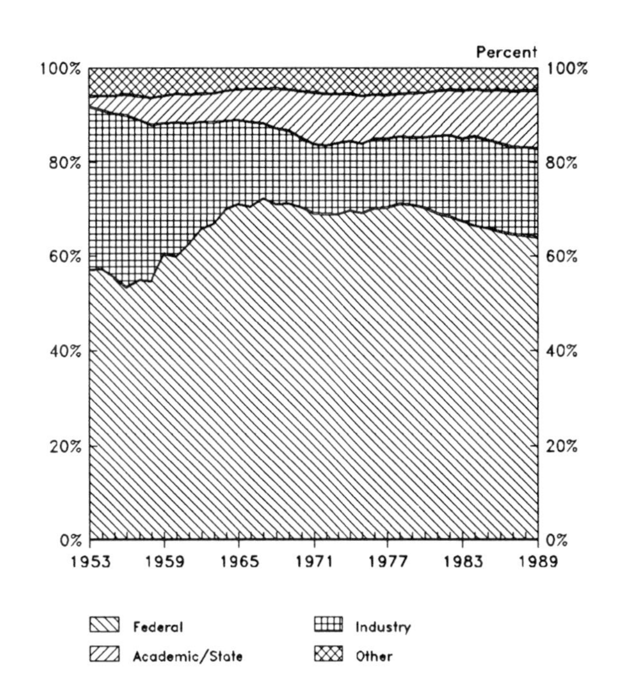

NOTE: Data series within the figures are not overlapped; top line represents total. Financial data are expressed in 1988 constant dollars to reflect real long-term growth trends.

DEFINITION OF TERMS: Basic research is a systematic study where the primary aim of the investigator is directed toward fuller knowledge or understanding of the subject under study, rather than a practical or commercial application thereof. Research expenditures include current fund expenditures for all research activities that are separately budgeted and accounted for. *Federal* sector consists of all agencies of the federal government. *Industry* sector consists of both manufacturing and non-manufacturing companies. Industry funding of industrial research includes all funds (e.g. state and local) other than those received from the federal government. *Academic/State* sector consists of all institutions of higher education, both public and private. Academic funding of research and development includes state and local government funds separately budgeted for R&D; and general educational funds, from any source, that the institutions have been free to allocate for separately budgeted research. *Other* sector consists of institutions that are primarily granting in nature, such as private philanthropic foundations and voluntary health agencies.

SOURCE: National Science Foundation, Division of Policy Research and Analysis. Database: CASPAR. Some of the data within this database are estimates, incorporated where there are discontinuities within data series or gaps in data collection. Primary data sources: National Science Foundation, Division of Science Resource Studies, Survey of Federal Funds for Research and Development; Survey of Federal Support to Universities, Colleges, and Non-profit Organizations; Survey of Scientific and Engineering Expenditures at Universities and Colleges; Survey of Industrial Research and Development.

ACADEMIC R&D EXPENDITURES

Academic R&D: Character of Research

Academic R&D expenditures increased steeply through the 1960s, from $2 billion (1988 dollars) in 1958 to $7 billion by 1968; they remained roughly level for a decade, then increased rapidly to $13 billion in 1988. The nature of academic research has shifted sharply since the early 1950s: Basic research increased from 45 percent of total academic R&D expenditures in 1953 to almost 80 percent in 1964; since the mid-1970s, however, it has fluctuated near 70 percent.

Figure 2-23: Academic R&D Expenditures by Type of R&D

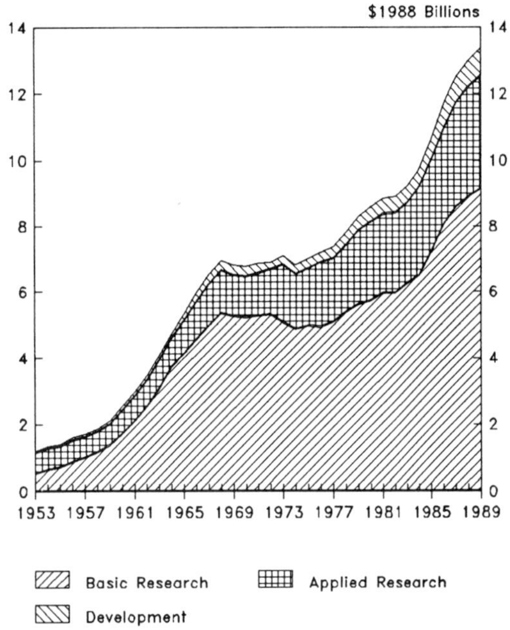

Figure 2-24: Distribution of Academic R&D Expenditures by Type of R&D

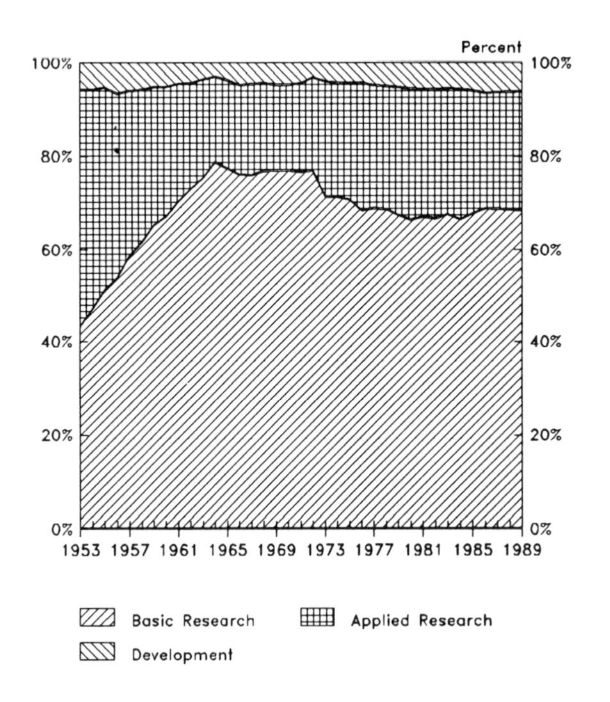

NOTE: Data series within the figures are not overlapped; top line represents total. Financial data are expressed in 1988 constant dollars to reflect real long-term growth trends.

DEFINITION OF TERMS: Academic R&D expenditures include current fund expenditures within higher education institutions for all research and development activities that are separately budgeted and accounted for. This includes both sponsored research activities (sponsored by federal and non-federal agencies and organizations) and university research separately budgeted under an internal application of institutional funds; but excludes training, public service, demonstration projects, departmental research not separately budgeted and FFRDCs. *Basic Research* is a systematic study where the primary aim of the investigator is directed toward fuller knowledge or understanding of the subject under study, rather than a practical or commercial application thereof. *Applied Research* is the systematic study where the primary aim of the investigator is directed toward gaining knowledge or understanding necessary for determining the means by which a recognized and specific need or commercial objective may be met. *Development* is the systematic use of the knowledge or understanding gained from research, directed toward the production of useful materials, devices, systems, or methods, including design and development of prototypes and processes.

SOURCE: National Science Foundation, Division of Policy Research and Analysis. Database: CASPAR. Some of the data within this database are estimates, incorporated where there are discontinuities within data series or gaps in data collection. Primary data source: National Science Foundation, Division of Science Resource Studies, Survey of Scientific and Engineering Expenditures at Universities and Colleges.

Academic R&D: Science and Engineering Fields

As a share of total academic R&D expenditures, the life sciences have increased from just over 40 percent in the late 1950s to about 55 percent by the mid-1970s. They have remained at 1970s levels throughout the 1980s. Conversely, the share of the physical sciences has declined from 20 percent in the 1950s to just above 10 percent by the mid-1970s, where it has remained. The share of the social and other behavioral sciences doubled from 7 percent in the late 1950s to 14 percent during the 1960s; during the 1980s, it returned to below 10 percent of the total.

Figure 2-25: Academic R&D Expenditures by Science and Engineering Field

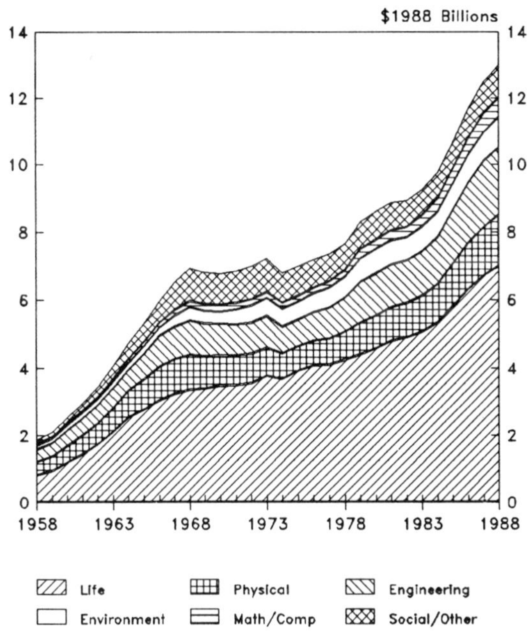

Figure 2-26: Distribution of Academic R&D Expenditures by Science and Engineering Field

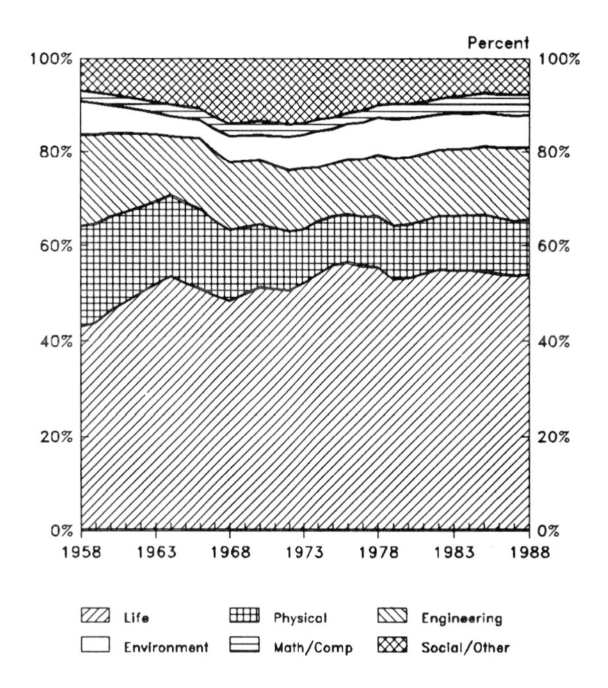

NOTE: Data series within the figures are not overlapped; top line represents total. Financial data are expressed in 1988 constant dollars to reflect real long-term growth trends.

DEFINITION OF TERMS: Academic R&D expenditures include current-fund expenditures within higher education institutions for all research and development activities that are separately budgeted and accounted for. This includes both sponsored research activities (sponsored by federal and non-federal agencies and organizations) and university research separately budgeted under an internal application of institutional funds; but excludes training, public service, demonstration projects, departmental research not separately budgeted, and FFRDCs. *Life* sciences include agricultural, biological, medical, and other health sciences. *Physical* sciences include astronomy, chemistry, and physics. *Engineering* includes aeronautical and astronautical, chemical, civil, electrical, and mechanical engineering. *Environment* includes oceanography, atmospheric, and earth sciences. *Mathematics/Computer* science includes all fields of mathematics and computer-related sciences. *Social/Other* include economics, political science, psychology, and sociology.

SOURCE: National Science Foundation, Division of Policy Research and Analysis. Database: CASPAR. Some of the data within this database are estimates, incorporated where there are discontinuities within data series or gaps in data collection. Primary data source: National Science Foundation, Division of Science Resource Studies, Survey of Scientific and Engineering Expenditures at Universities and Colleges.

Academic R&D: Source of Funding

The federal government's share of support for academic R&D increased from 55 percent in 1958 to more than 70 percent during the 1960s; from there it has gradually declined to its present level of 60 percent. The share contributed directly by the academic institutions themselves has increased from a little over 5 percent in the late 1950s to just under 20 percent in the late 1980s.

Figure 2-27: Academic R&D Expenditures by Source

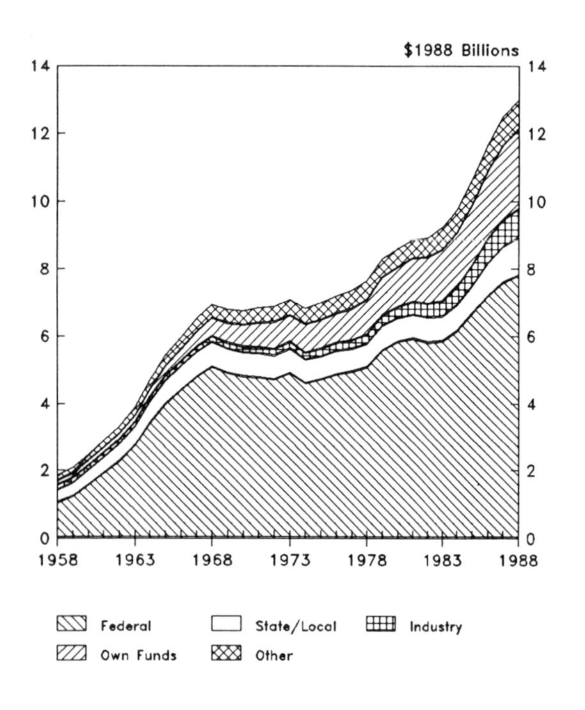

Figure 2-28: Distribution of Academic R&D Expenditures by Source

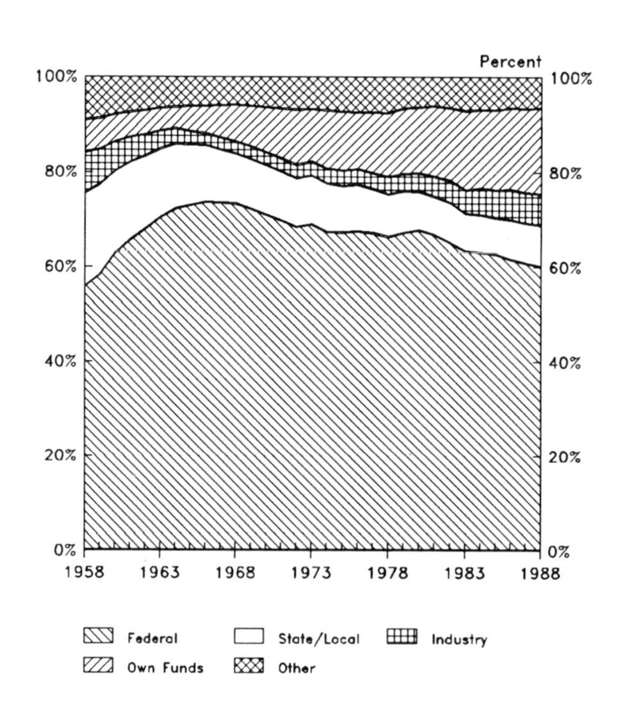

NOTE: Data series within the figures are not overlapped; top line represents total. Financial data are expressed in 1988 constant dollars to reflect real long-term growth trends.

DEFINITION OF TERMS: Academic R&D expenditures include current-fund expenditures within higher education institutions for all research and development activities that are separately budgeted and accounted for. This includes both sponsored research activities (sponsored by federal and non-federal agencies and organizations) and university research separately budgeted under an internal application of institutional funds; but excludes training, public service, demonstration projects, departmental research not separately budgeted, and FFRDCs. *Federal* funds include grants and contracts for academic R&D (including direct and reimbursed indirect costs) by agencies of the federal government. *State/Local* funds include funds for academic R&D from state, county, municipal, or other local governments and their agencies, including funds for R&D at agricultural and other experiment stations. *Industry* funds includes all grants and contracts for academic R&D from profit-making organizations, whether engaged in production, distribution, research, service, or other activities. *Own Funds* include institutional funds for separately budgeted research and development, cost-sharing, and under-recovery of indirect costs; they are derived from (1) general purpose state or local government appropriations, (2) general purpose grants from industry, foundations, and other outside sources, (3) tuition and fees, and (4) endowment income. *Other* sources include grants for academic R&D from non-profit foundations and voluntary health agencies, as well as individual gifts that are restricted by the donor to research.

SOURCE: National Science Foundation, Division of Policy Research and Analysis. Database: CASPAR. Some of the data within this database are estimates, incorporated where there are discontinuities within data series or gaps in data collection. Primary data source: National Science Foundation, Division of Science Resource Studies, Survey of Scientific and Engineering Expenditures at Universities and Colleges.

Academic R&D: Type of Institutions

For the past two decades, doctoral institutions have maintained a 90-percent share of all academic R&D expenditures. In 1988, doctoral institution R&D expenditures totalled $11.5 billion.

Figure 2-29: Academic R&D Expenditures by Institution Type

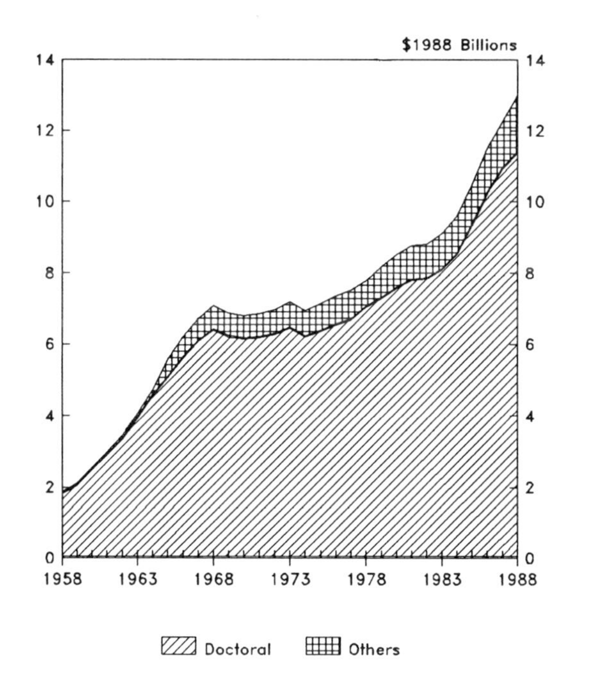

Figure 2-30: Distribution of Academic R&D Expenditures by Institution Type

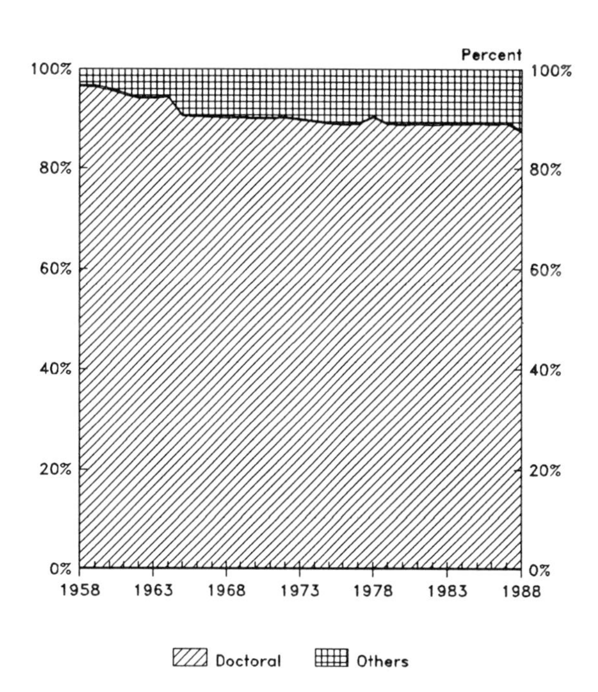

NOTE: Data series within the figures are not overlapped; top line represents total. Financial data are expressed in 1988 constant dollars to reflect real long-term growth trends.

DEFINITION OF TERMS: Academic R&D expenditures include current-fund expenditures within higher education institutions for all research and development activities that are separately budgeted and accounted for. This includes both sponsored research activities (sponsored by federal and non-federal agencies and organizations) and university research separately budgeted under an internal application of institutional funds; but excludes training, public service, demonstration projects, departmental research not separately budgeted, and FFRDCs. *Doctoral* institutions are higher education institutions that have granted an average of 10 or more Ph.D. degrees per year in the natural sciences or engineering over the past two decades; they include 116 public and 69 private institutions. *Other* includes comprehensive institutions that grant at least half of their degrees for courses of study that normally require 4 or more years to complete, and 2-year institutions that award primarily 2-year associate or technician degrees.

SOURCE: National Science Foundation, Division of Policy Research and Analysis. Database: CASPAR. Some of the data within this database are estimates, incorporated where there are discontinuities within data series or gaps in data collection. Primary data source: National Science Foundation, Division of Science Resource Studies, Survey of Scientific and Engineering Expenditures at Universities and Colleges.

Sources of R&D Funding: Private Doctoral Institutions

Private doctoral institution R&D expenditures increased from $750 thousand (1988 dollars) in 1958 to $2.5 billion in 1968. After remaining roughly level during the 1970s they climbed to nearly $4 billion by 1988. The federal share of support increased from 66 percent in 1958 to over 80 percent during the 1960s; since then it has gradually declined to its 1988 level of 73 percent. The share contributed directly by the institutions has increased from 3 percent in the late 1950s to nearly 10 percent in the late 1980s.

Figure 2-31: Private Doctoral Institution R&D Expenditures by Source of Funds

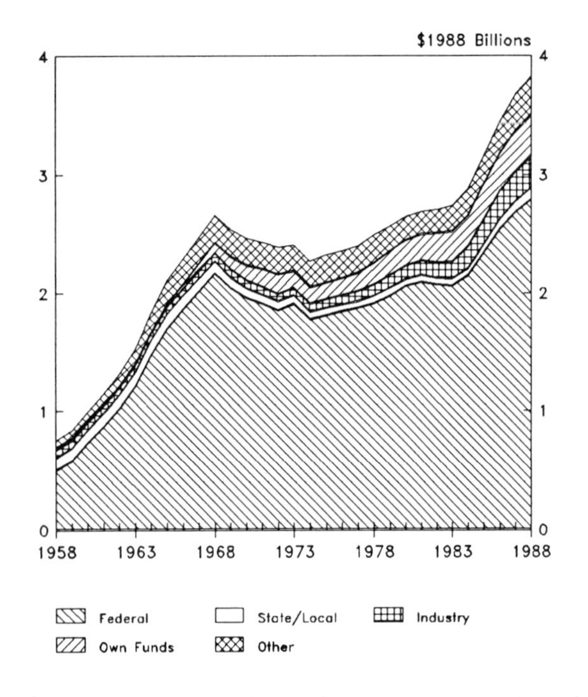

Figure 2-32: Distribution of Private Doctoral Institution R&D Expenditures by Source of Funds

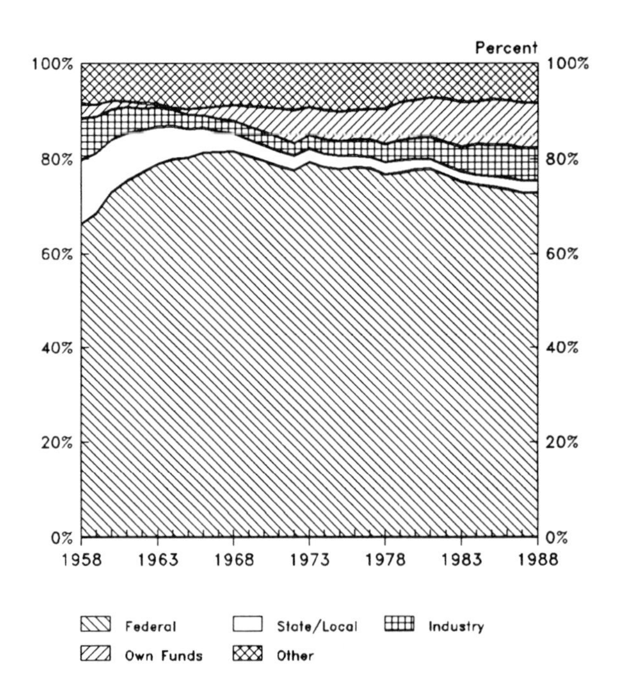

NOTE: Data series within the figures are not overlapped; top line represents total. Financial data are expressed in 1988 constant dollars to reflect real long-term growth trends.

DEFINITION OF TERMS: Private doctoral institutions are institutions that have granted an average of 10 or more Ph.D. degrees per year in the natural sciences or engineering over the past two decades, and are under the control of--or affiliated with--non-profit, independent organizations with or without religious affiliation; they include 69 institutions. R&D expenditures include current-fund expenditures within doctoral institutions for all research and development activities that are separately budgeted and accounted for; excluding departmental research not separately budgeted and FFRDCs. *Federal* funds include grants and contracts for R&D (including direct and reimbursed indirect costs) by agencies of the federal government, excluding funds for FFRDCs. *State/Local* funds include funds for R&D from state, county, municipal, or other local governments and their agencies, including funds for R&D at agricultural and other experiment stations. *Industry* funds include all grants and contracts for R&D from profit-making organizations, whether engaged in production, distribution, research, service, or other activities. *Own Funds* include institutional funds for separately budgeted research and development, cost-sharing, and under-recovery of indirect costs. They are derived from (1) general purpose state or local government appropriations, (2) general purpose grants from industry, foundations, or other outside sources, (3) tuition and fees, and (4) endowment income. *Other* sources include grants for R&D from non-profit foundations and voluntary health agencies, as well as individual gifts that are restricted by the donor to research.

SOURCE: National Science Foundation, Division of Policy Research and Analysis. Database: CASPAR. Some of the data within this database are estimates, incorporated where there are discontinuities within data series or gaps in data collection. Primary data source: National Science Foundation, Division of Science Resource Studies, Survey of Scientific and Engineering Expenditures at Universities and Colleges.

Sources of R&D Funding: Public Doctoral Institutions

Public doctoral institution R&D expenditures increased through the 1960s from over $1 billion (1988 dollars) in 1958 to $4 billion in 1968. After remaining roughly level during the 1970s, they have climbed to more than $7.5 billion by 1988. The federal government's share of support increased from 48 percent in 1958 to 68 percent in 1968, then gradually declined to its 1988 level of 53 percent. The share contributed directly by the institutions has increased from 10 percent in the late 1950s to over 20 percent in the late 1980s.

Figure 2-33: Public Doctoral Institution R&D Expenditures by Source of Funds

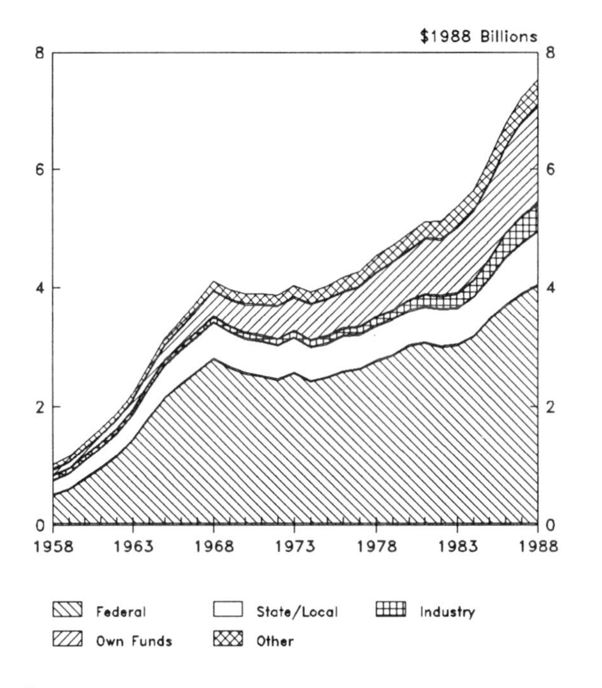

Figure 2-34: Distribution of Public Doctoral Institution R&D Expenditures by Source of Funds

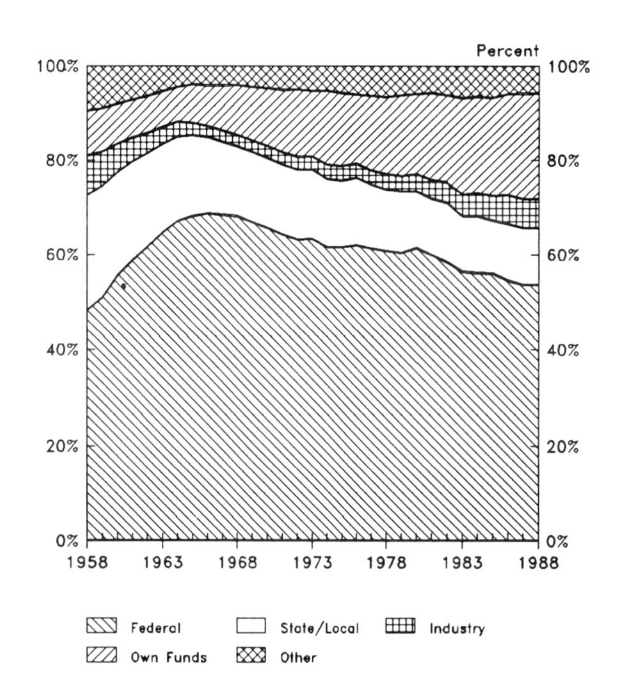

NOTE: Data series within the figures are not overlapped; top line represents total. Financial data are expressed in 1988 constant dollars to reflect real long-term growth trends.

DEFINITION OF TERMS: Public doctoral institutions are institutions that have granted an average of 10 or more Ph.D. degrees per year in the natural sciences or engineering over the past two decades, and are under the control of--or affiliated with--federal, state, local, state and local, or state-related agencies; they include 116 institutions. R&D Expenditures include current-fund expenditures within doctoral institutions for all research and development activities that are separately budgeted and accounted for; excluding departmental research not separately budgeted and FFRDCs. *Federal* funds include grants and contracts for R&D (including direct and reimbursed indirect costs) by agencies of the federal government, excluding funds for FFRDCs. *State/Local* funds include funds for R&D from state, county, municipal, or other local governments and their agencies, including funds for R&D at agricultural and other experiment stations. *Industry* funds include all grants and contracts for R&D from profit-making organizations, whether engaged in production, distribution, research, service, or other activities. *Own Funds* include institutional funds for separately budgeted research and development, cost-sharing, and under-recovery of indirect costs. They are derived from (1) general purpose state or local government appropriations, (2) general purpose grants from industry, foundations, or other outside sources, (3) tuition and fees, and (4) endowment income. *Other* sources include grants for R&D from non-profit foundations and voluntary health agencies, as well as individual gifts that are restricted by the donor to research.

SOURCE: National Science Foundation, Division of Policy Research and Analysis. Database: CASPAR. Some of the data within this database are estimates, incorporated where there are discontinuities within data series or gaps in data collection. Primary data source: National Science Foundation, Division of Science Resource Studies, Survey of Scientific and Engineering Expenditures at Universities and Colleges.

Academic S&E Facilities: Research and Instruction

Annual expenditures for academic research facilities increased from an estimated $0.3 billion (1988 dollars) in 1958 to about $1 billion in 1968, declined through the 1970s, then increased to more than $1 billion in the late 1980s. The share of academic science and engineering facilities expenditures for research purposes is estimated to have risen to 60 percent by the late 1980s.

Figure 2-35: Academic Expenditures for Science and Engineering Facilities by Purpose

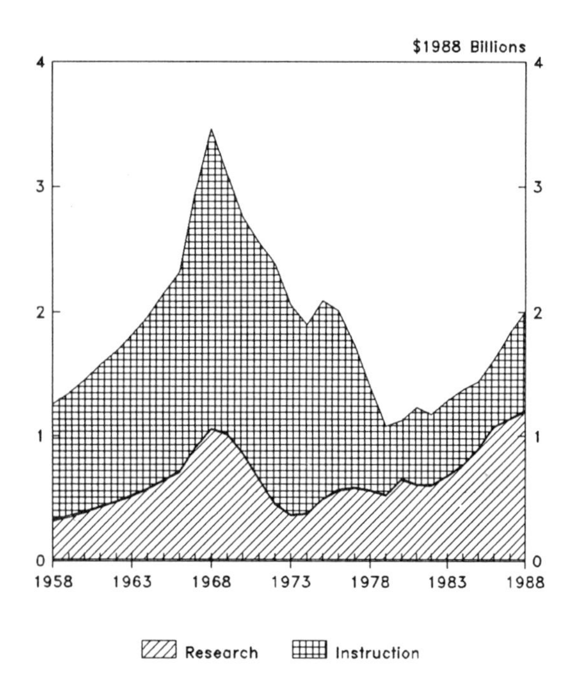

Figure 2-36: Distribution of Academic Expenditures for Science and Engineering Facilities by Purpose

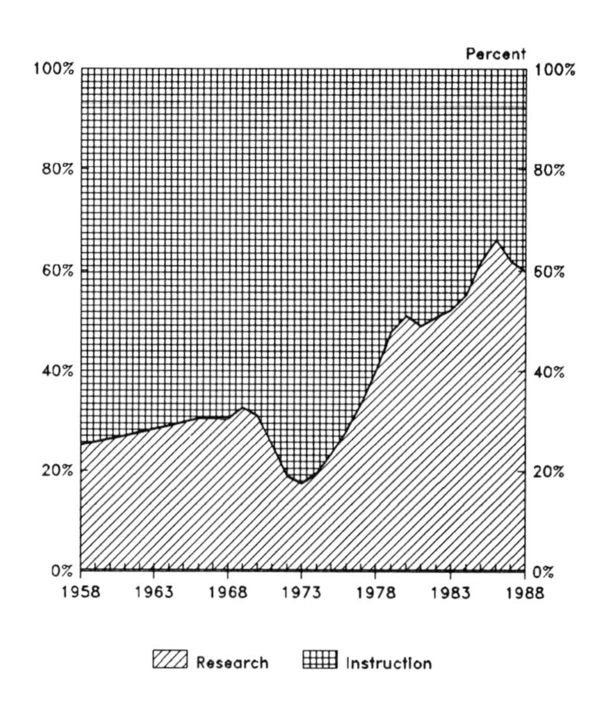

NOTE: Data series within the figures are not overlapped; top line represents total. Financial data are expressed in 1988 constant dollars to reflect real long-term growth trends.

DEFINITION OF TERMS: Academic science and engineering facilities expenditures include estimated capital expenditures for research and instructional facilities including fixed or built-in equipment; some movable equipment and movable furnishings, such as desks; and facilities constructed to house scientific apparatus. Expenditure shares attributed to *Research* and *Instruction* purposes are estimates based on undergraduate and graduate enrollment data, as well as data on faculty positions assigned to teaching and research.

SOURCE: National Science Foundation, Division of Policy Research and Analysis. Database: CASPAR. Some of the data within this database are estimates, incorporated where there are discontinuities within data series or gaps in data collection. Primary data source: National Science Foundation, Division of Science Resource Studies, Survey of Scientific and Engineering Expenditures at Universities and Colleges; U.S. Department of Education National Center for Education Statistics, Higher Education General Survey (HEGIS): Fall Enrollment in Institution of Higher Education.

Academic S&E Facilities: Source of Funding

Annual capital expenditures for academic science and engineering facilities (for both research and instruction) increased sharply from $1.3 billion (1988 dollars) in 1958 to $3.5 billion in 1968, declined sharply to $1 billion in 1979, then rose to $2 billion in 1988. The federal share of these funds increased from 27 percent in 1958 to 32 percent in the 1960s, then declined to its present level of 11 percent.

Figure 2-37: Expenditures for Academic Science and Engineering Facilities by Source of Funds

$1988 Billions
4
3
2
1
0
1958
1963
1968
1973
1978
1983
1988
Federal
Other

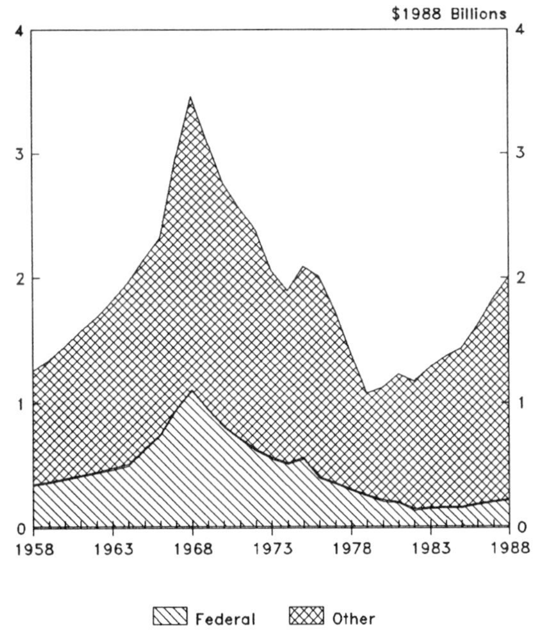

Figure 2-38: Distribution of Expenditures for Academic Science and Engineering Facilities by Source of Funds

Percent
100%
80%
60%
40%
20%
0%
1958
1963
1968
1973
1978
1983
1988
Federal
Other

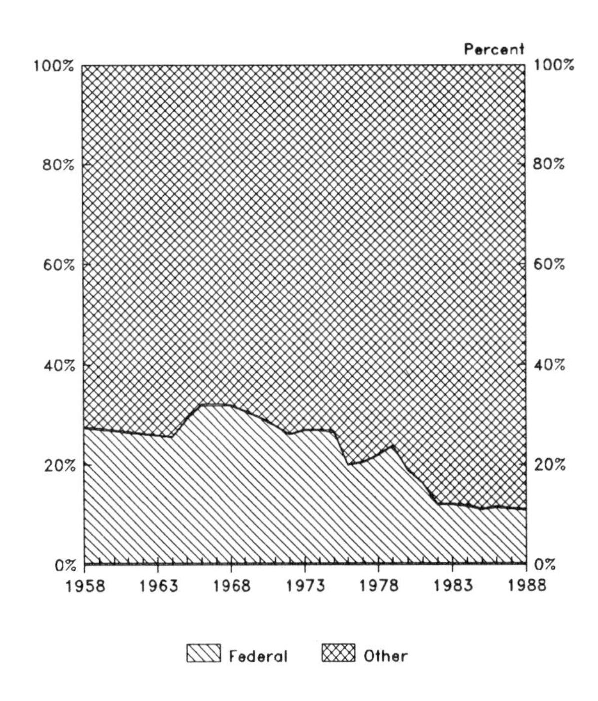

NOTE: Data series within the figures are not overlapped; top line represents total. Financial data are expressed in 1988 constant dollars to reflect real long-term growth trends.

DEFINITION OF TERMS: Academic science and engineering facilities expenditures include capital expenditures for research and instructional facilities, including fixed or built-in equipment, some movable equipment and movable furnishings such as desks, and facilities constructed to house scientific apparatus. *Federal* funds include expenditures for academic science and engineering facilities with moneys from federal agency contracts in grants. *Other* sources include state and local governments, the institutions themselves, industry, and other non-profit organizations.

SOURCE: National Science Foundation, Division of Policy Research and Analysis. Database: CASPAR. Some of the data within this database are estimates, incorporated where there are discontinuities within data series or gaps in data collection. Primary data source: National Science Foundation, Division of Science Resource Studies, Survey of Scientific and Engineering Expenditures at Universities and Colleges.

Academic Research Equipment: Source of Funds

Expenditures for academic research equipment have increased from less than $200 million (1988 dollars) in 1958 to $600 million in the mid-1960s; they fell during the 1970s, but have increased substantially in the 1980s to nearly $900 million. The federal share of academic research equipment funds has declined from roughly 75 percent in 1958 to about 60 percent in the late 1980s.

Figure 2-39: Expenditures for Academic Research Equipment by Source of Funds

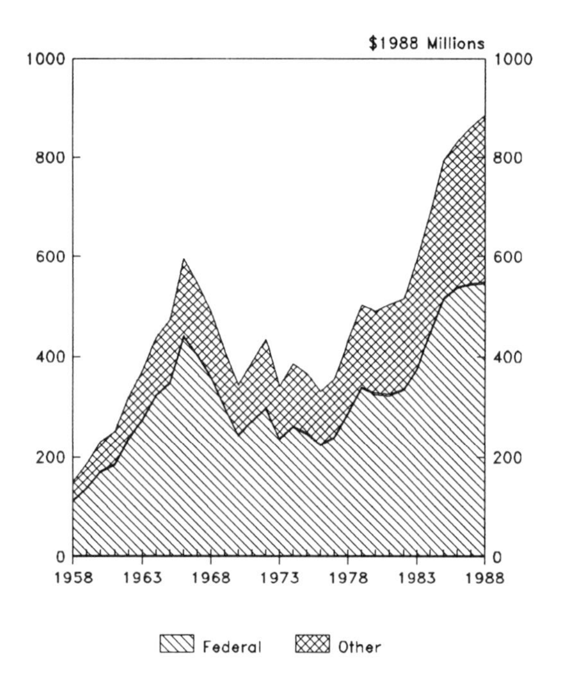

Figure 2-40: Distribution of Expenditures for Academic Research Equipment by Source of Funds

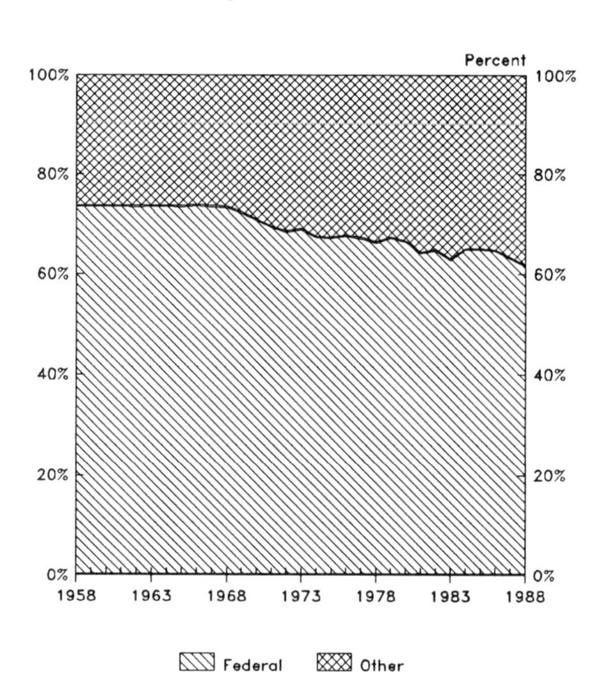

NOTE: Data series within the figures are not overlapped; top line represents total. Financial data are expressed in 1988 constant dollars to reflect real long-term growth trends.

DEFINITION OF TERMS: Research equipment expenditures include (1) reported expenditures of separately budgeted current-funds for the purchase of research equipment, and (2) estimated capital expenditures for fixed or built-in research equipment and furniture. *Federal* funds include expenditures for academic research equipment with monies from grants and contracts for academic R&D (including direct and reimbursed indirect costs) by agencies of the federal government; excludes expenditures for FFRDC facilities. *Other* sources include state and local governments, the institution themselves, industry, and other non-profit organizations.

SOURCE: National Science Foundation, Division of Policy Research and Analysis. Database: CASPAR. Some of the data within this database are estimates, incorporated where there are discontinuities within data series or gaps in data collection. Primary data source: National Science Foundation, Division of Science Resource Studies, Survey of Scientific and Engineering Expenditures at Universities and Colleges.

Academic R&D Expenditures per Investigator: Equipment and Facilities

Estimated expenditures for academic R&D equipment per investigator more than doubled during the 1960s, accounting for inflation, from $5,500 (1988 dollars) in 1958 to $13,500 in 1966, falling to $6,400 by 1974, then rising rapidly to nearly $13,000 by 1988. Similarly, estimated expenditures for academic R&D facilities per investigator doubled during the 1960s, from $11,600 in 1958 to $21,400 in 1968, then plummeted to $7,600 in 1974, rising rapidly again during the 1980s to $17,600 by 1988.

Figure 2-41: Academic Expenditures for R&D Equipment per FTE Investigator

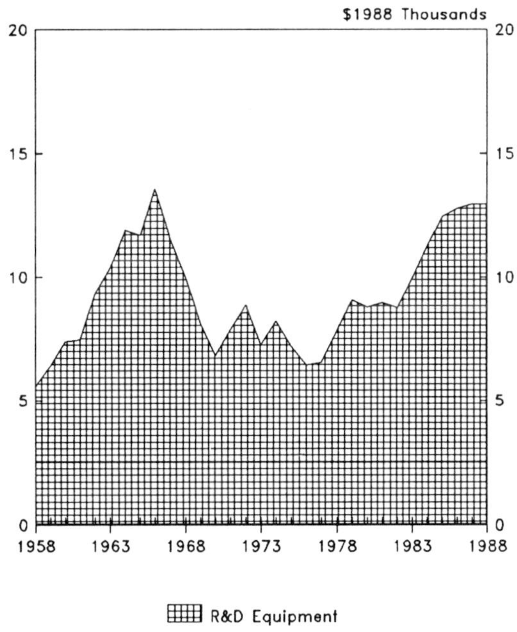

$1988 Thousands
20
15
10
5
0
1958 1963 1968 1973 1978 1983 1988
R&D Equipment

Figure 2-42: Academic Expenditures for R&D Facilities per FTE Investigator

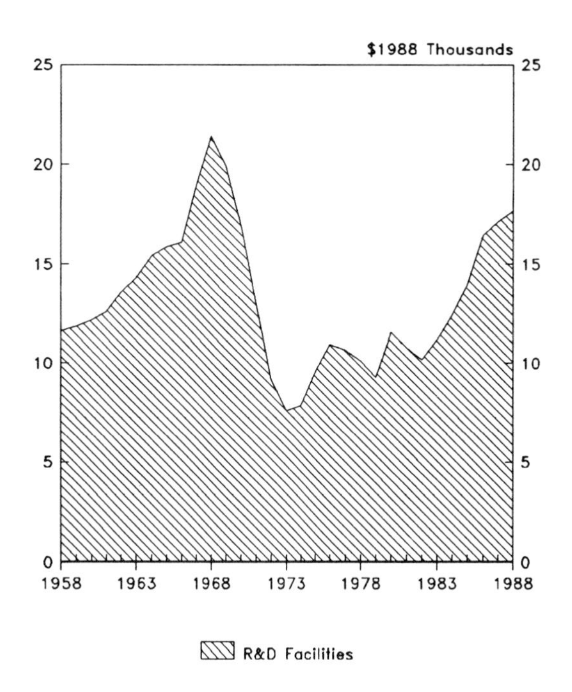

NOTE: Financial data are expressed in 1988 constant dollars to reflect real long-term growth trends.

DEFINITION OF TERMS: *R&D Equipment* expenditures include (1) reported expenditures of separately budgeted current-funds for the purchase of research equipment, and (2) estimated capital expenditures for fixed or built-in research equipment. *R&D Facilities* expenditures include estimated capital expenditures for research facilities. Facilities expenditures are estimated shares of reported expenditures for academic science and engineering facilities; based on undergraduate and graduate enrollment data, as well as faculty positions assigned to research and teaching. FTE (full-time equivalent) investigators include scientists and engineers conducting funded (separately budgeted) academic R&D; the full-time equivalent is an estimate derived from the fraction of faculty time spent in those research activities, non-faculty scientists and engineers employed to conduct research in campus facilities (except FFRDCs), and post-doctoral researchers working in academic institutions.

SOURCE: National Science Foundation, Division of Policy Research and Analysis. Database: CASPAR. Some of the data within this database are estimates, incorporated where there are discontinuities within data series or gaps in data collection. Primary data source: National Science Foundation, Division of Science Resource Studies, Survey of Scientific and Engineering Expenditures at Universities and Colleges, Survey of Scientific and Engineering Personnel Employed at Universities and Colleges.

Total Academic R&D: Estimated Cost Components

For the past three decades, personnel costs have accounted for about 45 percent of total costs related to the conduct of academic research, with 40 percent supporting senior personnel and 5 percent supporting graduate students. Other direct costs have fluctuated between 15 percent and 20 percent. The indirect-cost share doubled from 15 percent in the 1950s to nearly 30 percent in 1980, where it has steadied. The combined share for equipment and facilities declined from over 20 percent through the 1960s to 10 percent in the 1970s; it has since increased to 15 percent.

Figure 2-43: Estimated Cost Components of U.S. Academic R&D Budget

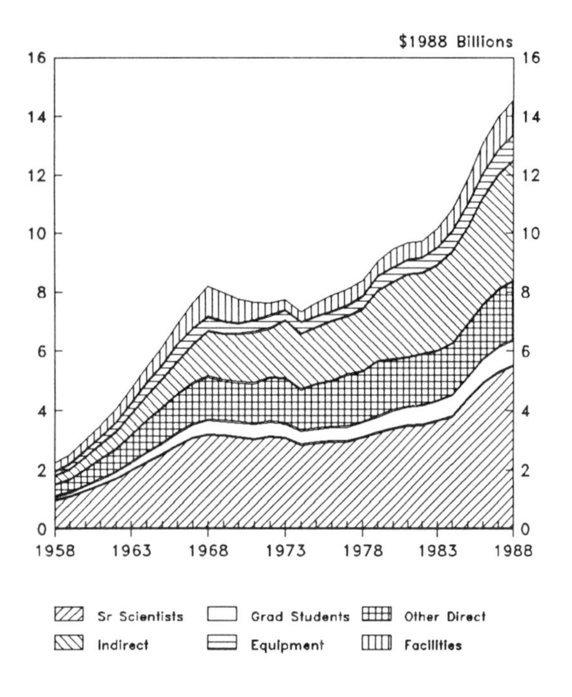

Figure 2-44: Distribution of Estimated Cost Components of U.S. Academic R&D Budgets

NOTE: Data series within the figures are not overlapped; top line represents total. Financial data are expressed in 1988 constant dollars to reflect real long-term growth trends.

DEFINITION OF TERMS: Estimated personnel costs for *Senior Scientists* and *Graduate Students* include salaries and fringe benefits, such as insurance and retirement contributions. *Other Direct* costs include such budget items as materials and supplies, travel, subcontractors, computer services, publications, consultants, and participant support costs. *Indirect* costs include general administration, department administration, building operation and maintenance, depreciation and use, sponsored-research projects administration, libraries, and student-services administration. *Equipment* costs include (1) reported expenditures of separately budgeted current-funds for the purchase of research equipment, and (2) estimated capital expenditures for fixed or built-in research equipment. *Facilities* costs include estimated capital expenditures for research facilities, including facilities constructed to house scientific apparatus.

SOURCE: National Science Foundation, Division of Policy Research and Analysis. Database: CASPAR. Some of the data within this database are estimates, incorporated where there are discontinuities within data series or gaps in data collection. Primary data sources: National Science Foundation, Division of Science Resource Studies; Survey of Scientific and Engineering Expenditures at Universities and Colleges: National Institutes of Health; American Association of University Professors; National Association of State Universities and Land Grant Colleges.

Total Academic R&D: Estimated Expenditures Per Investigator

Annual expenditures--including operating, equipment, and capital spending--per academic investigator (FTE) are estimated to have increased from $85,000 (1988 dollars) in 1958 to about $170,000 by the late 1960s, where they leveled off for a decade; in the 1980s, they increased again to $225,000.

Figure 2-45: Academic R&D Expenditures per FTE Investigator by Type of Expenditure

Figure 2-46: Distribution of Academic R&D Expenditures per FTE Investigator by Type of Expenditure

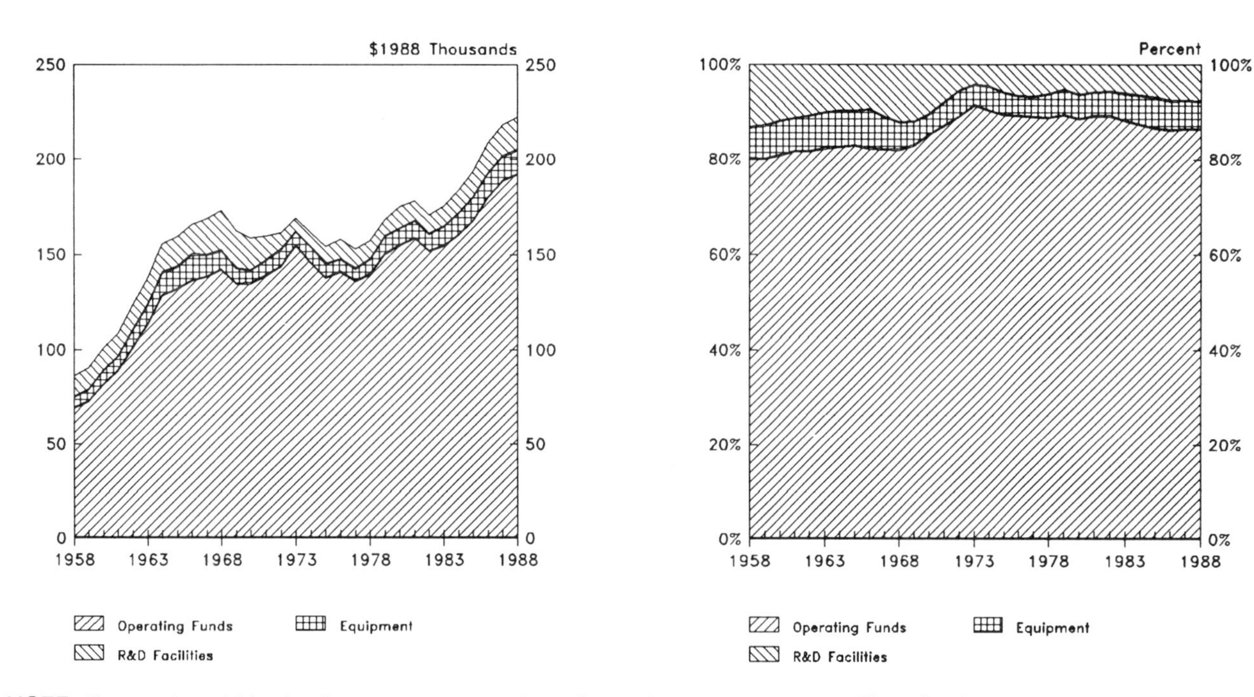

NOTE: Data series within the figures are not overlapped; top line represents total. Financial data are expressed in 1988 constant dollars to reflect real long-term growth trends.

DEFINITION OF TERMS: *Operating Funds* include current-fund expenditures for academic research and development activities that are separately budgeted and accounted for; includes expenditures for senior scientist and graduate student compensation, other direct costs, and indirect costs associated with conduct of academic research. *Equipment* includes (1) reported expenditures of separately budgeted current-funds for the purchase of academic research equipment, and (2) estimated capital expenditures for fixed or built-in research equipment. *R&D Facilities* include estimated capital expenditures for academic research facilities. *FTE Investigators* include those scientists and engineers conducting funded (separately budgeted) academic R&D; the full-time equivalent is an estimate, derived from the fraction of faculty time spent in those research activities, non-faculty scientists and engineers employed to conduct research in campus facilities (except FFRDCs), and post-doctoral researchers working in academic institutions.

SOURCE: National Science Foundation, Division of Policy Research and Analysis. Database: CASPAR. Some of the data within this database are estimates, incorporated where there are discontinuities within data series or gaps in data collection. Primary data source: National Science Foundation, Division of Science Resource Studies, Survey of Scientific and Engineering Expenditures at Universities and Colleges, Survey of Scientific and Engineering Personnel Employed at Universities and Colleges.

Personnel Expenditures: Natural Sciences and Engineering

After a decade of slow decline, accounting for inflation, the average total compensation for academic Ph.D.s in the natural sciences and engineering increased during the 1980s, from $59,000 ($1988 dollars) in 1981 to more than $70,000 in 1988.

Figure 2-47: Average Salary and Benefits Paid Academic Ph.D.s in Natural Sciences and Engineering

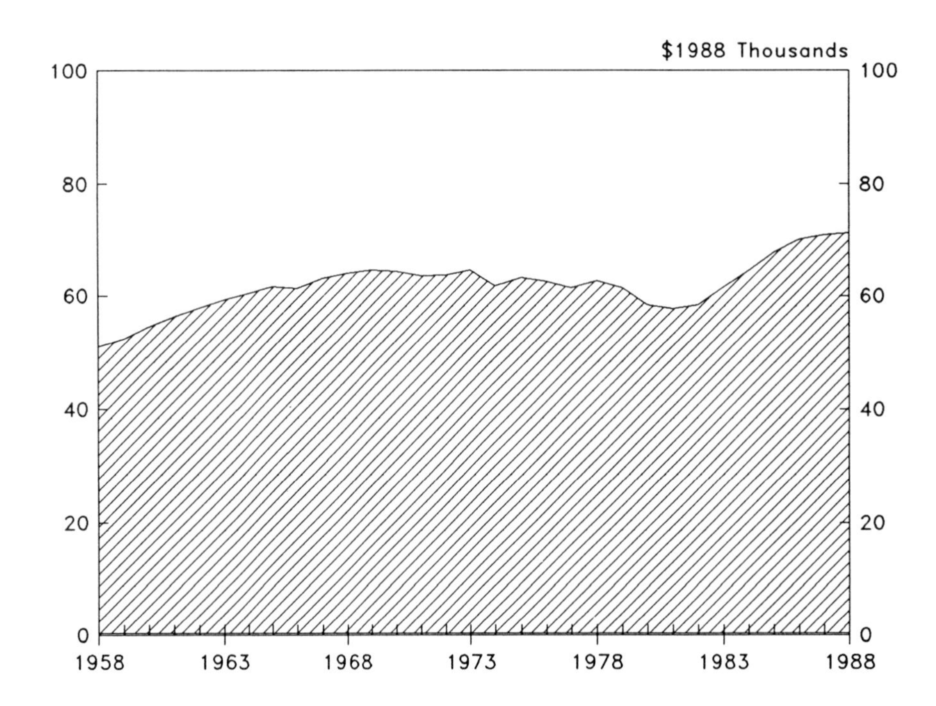

NOTE: Financial data are expressed in 1988 constant dollars to reflect real long-term growth trends.

DEFINITION OF TERMS: Academic Ph.D.s in the natural sciences and engineering include academic employees who have been awarded the Ph.D. degree within the following fields: life sciences, including agricultural, biological, medical, and other health sciences; physical sciences, including astronomy, chemistry, and physics; engineering, including aeronautical and astronautical, chemical, civil, electrical, and mechanical engineering; environmental sciences, including oceanography, atmospheric and earth sciences; mathematics and computer science, including all fields of mathematics and computer-related sciences. Compensation includes salaries and fringe benefits, including insurance and retirement contributions.

SOURCE: National Science Foundation, Division of Policy Research and Analysis. Database: CASPAR. Some of the data within this database are estimates, incorporated where there are discontinuities within data series or gaps in data collection. Primary data sources: National Science Foundation, Division of Science Resource Studies, Survey of Scientific and Engineering Expenditures at Universities and Colleges, Survey of Scientific and Engineering Personnel Employed at Universities and Colleges; American Council on Education; National Association of State Universities and Grant Colleges.

TOTAL ACADEMIC EXPENDITURES AND REVENUES

Total Academic Operating Expenditures: Purpose

For the past 2 decades, over-all academic expenditure patterns have remained generally stable, with research accounting for 10 percent to 15 percent; education-related activities, 60 percent to 65 percent; public service, less than 5 percent; and other operations--hospitals, self-financing enterprises such as bookstores and dormitories, and federally financed research and development centers--accounting for more than 20 percent. Total national academic operating expenditures reached $110 billion in 1988.

Figure 2-48: Total Academic Operating Expenditures by Purpose

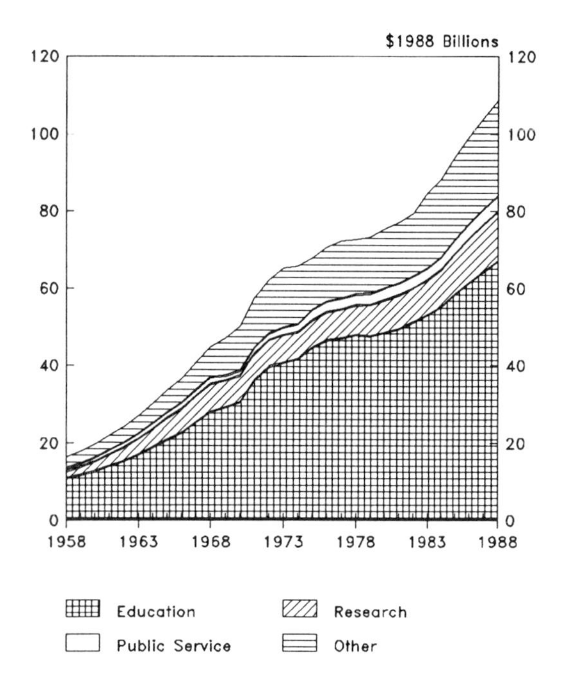

Figure 2-49: Distribution of Total Academic Operating Expenditures by Purpose

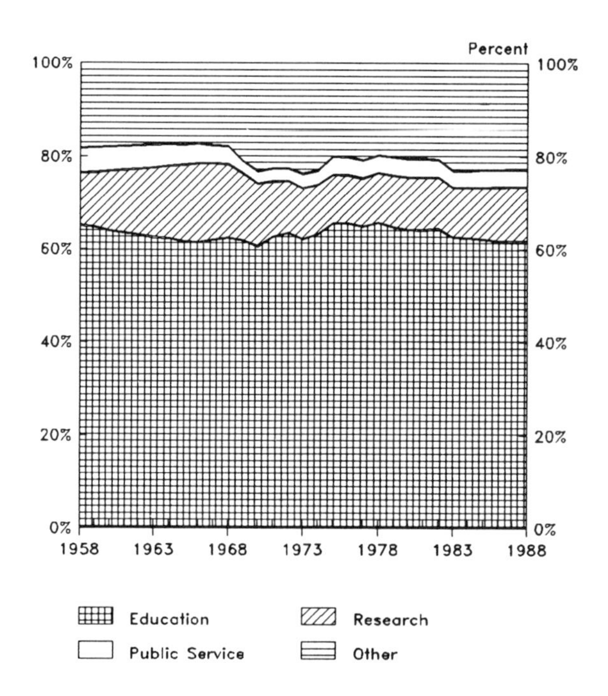

NOTE: Data series within the figures are not overlapped; top line represents total. Financial data are expressed in 1988 constant dollars to reflect real long-term growth trends.

DEFINITION OF TERMS: Academic institutions include 185 doctoral institutions, 1,224 comprehensive institutions, and 1,388 2-year institutions, the latter of which award primarily 2-year associate or technician degrees. Operating expenditures consist of educational and general current-fund expenditures for instruction, research, public service, academic support, student services, institutional support, operation and maintenance of plant, scholarships and fellowships, and educational and mandatory transfers for debt service; and for auxiliary enterprises and federally funded research and development centers, but exclude expenditures from institutional plant fund accounts. *Education* includes instructional expenditures, including departmental research not separately budgeted; current operating expenditures for libraries, operation and maintenance of plant, scholarships and fellowships, and student services. *Research* includes current fund expenditures for separately budgeted research and development. *Public Service* includes funds budgeted specifically for non-instructional services beneficial to groups external to the institution. *Other* includes hospitals, auxiliary enterprises, and (FFRDCs) administered by universities.

SOURCE: National Science Foundation, Division of Policy Research and Analysis. Database: CASPAR. Some of the data within this database are estimates, incorporated where there are discontinuities within data series or gaps in data collection. Primary data source: U.S. Department of Education, National Center for Education Statistics, Higher Education General Information Survey (HEGIS): Financial Statistics of Institutions of Higher Education.

Total Academic Operating Revenues: Sources

Over the past three decades, total academic operating revenues have increased more than 400 percent, accounting for inflation, exceeding $110 billion in 1988. This represents an average real growth rate of 6.4 percent per year since 1958. State funds and income from tuition contributed most to this growth; total federal funds to academic institutions have essentially been level since the late 1960s and, consequently, have declined as a share of academic revenues from 20 percent in 1968 to 10 percent in 1988.

Figure 2-50: Academic Institution Operating Revenues by Source of Funds

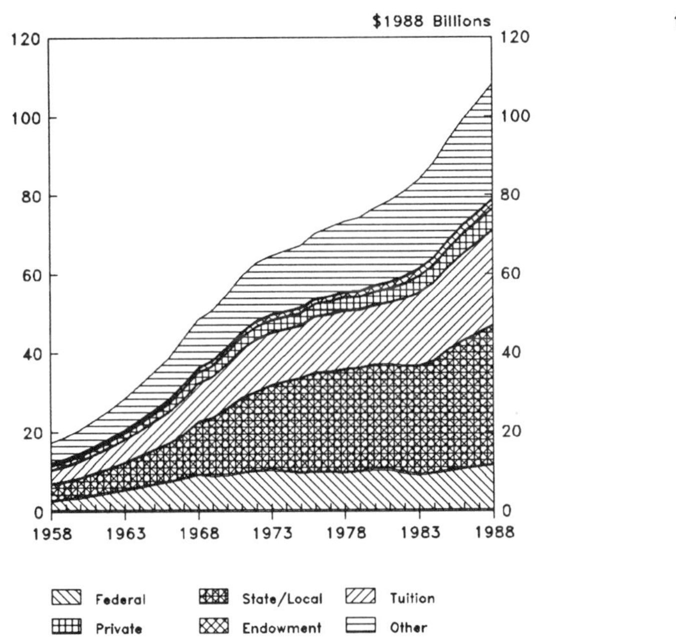

Figure 2-51: Distribution of Academic Institution Operating Revenues by Source of Funds

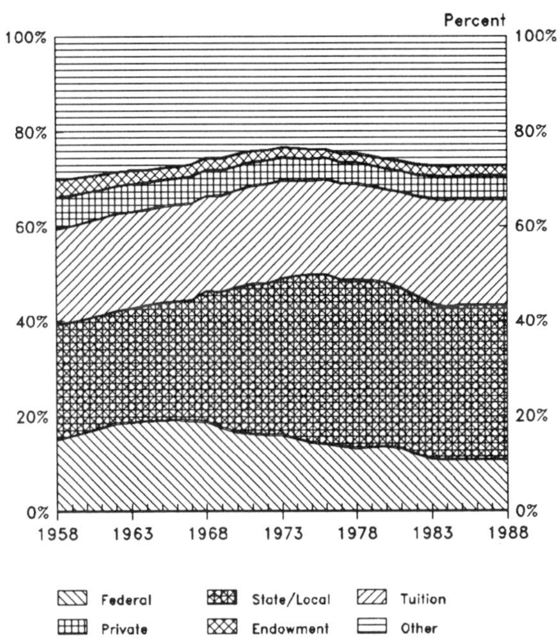

NOTE: Data series within the figures are not overlapped; top line represents total. Financial data are expressed in 1988 constant dollars to reflect real long-term growth trends.

DEFINITION OF TERMS: Academic institutions include 185 doctoral institutions, 1,224 comprehensive institutions, and 1,388 2-year institutions. *Federal* sources include (1) dollars appropriated or made available by the federal government to public or private institutions of higher education for current operating expenses, such as land-grant appropriations and revenue sharing funds and; grants and contracts for specific research projects; and other types of programs, such as administrative allowances for student aid; excludes funding for federally funded research and development centers (FRRDCs). *Tuition* include all assessments against students for current operating purposes, but charges for room, board, and other services rendered by auxiliary enterprises are not included. *State/Local* sources include dollars appropriated or made available by state and local governments to public or private institutions of higher education for current operating expenses and or for specific projects or programs. *Private* income includes private gifts and grants that are directly related to instruction, research, or public service; moneys received as a result of gifts, grants, or contracts from a foreign government are included, as well as an estimated dollar amount for contributed services. *Endowment* income includes the unrestricted income of endowment and similar funds. *Other* includes sales and services fees from educational activities; revenues derived from the sales of goods or services, and revenues from hospitals and FFRDCs.

SOURCE: National Science Foundation, Division of Policy Research and Analysis. Database: CASPAR. Some of the data within this database are estimates, incorporated where there are discontinuities within data series or gaps in data collection. Primary data source: U.S. Department of Education, National Center for Education Statistics, Higher Education General Information Survey (HEGIS): Financial Statistics of Institutions of Higher Education.

Doctoral Institution Operating Expenditures: Purpose

Total operating expenditures for doctoral institutions exceeded $65 billion in 1988, maintaining a 60-percent share of total academic expenditures since 1973. For the past two decades, over-all expenditure patterns of doctoral institutions have remained generally stable. Of total operating expenditures, research accounts for almost 20 percent, with education-related activities accounting for 50 percent, and other operations--hospitals, self-financing enterprises (such as bookstores and dormitories), and federally funded research and development centers (FFRDCs)--accounting for 25 percent.

Figure 2-52: Doctoral Institution Operating Expenditures by Purpose

Figure 2-53: Distribution of Doctoral Institution Operating Expenditures by Purpose

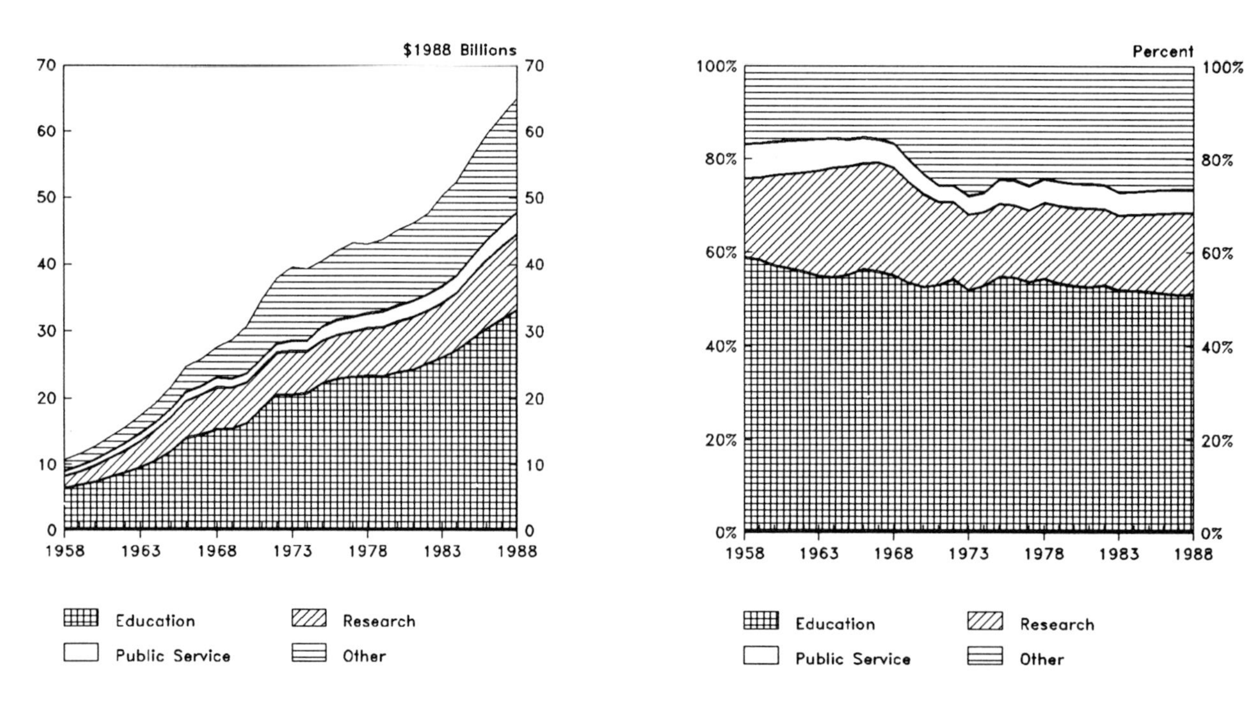

NOTE: Data series within the figures are not overlapped; top line represents total. Financial data are expressed in 1988 constant dollars to reflect real long-term growth trends.

DEFINITION OF TERMS: Doctoral institutions are higher education institutions that have granted an average of 10 or more Ph.D. degrees per year in the natural sciences or engineering over the past two decades; they include 116 public and 69 private institutions. *Education* includes instructional expenditures, including departmental research not separately budgeted; current operating expenditures for libraries, operation and maintenance of plant; scholarships and fellowships; and student services. *Research* includes current-fund expenditures for separately budgeted research and development. *Public Service* includes funds budgeted specifically for non-instructional services beneficial to groups external to the institution. *Other* includes hospitals, auxiliary enterprises, and federally funded research and development centers (FFRDCs) administered by universities.

SOURCE: National Science Foundation, Division of Policy Research and Analysis. Database: CASPAR. Some of the data within this database are estimates, incorporated where there are discontinuities within data series or gaps in data collection. Primary data source: U.S. Department of Education, National Center for Education Statistics, Higher Education General Information Survey (HEGIS): Financial Statistics of Institutions of Higher Education.

Doctoral Institution Expenditures: Per Faculty and Student

While growth in doctoral institution faculty and enrollments have slowed during the past decade, total expenditures of doctoral institutions have continued to rise. From 1978 to 1988, accounting for inflation, operational expenditures per faculty member have risen by more than 40 percent, reaching $260,000 in 1988. Education expenditures per student rose more than 30 percent between 1978 and 1988, reaching $9,500 in 1988.

Figure 2-54: Doctoral Institution Operating Expenditures per Faculty Member

$1988 Thousands

300
250
200
150
100
50
0
1958 1963 1968 1973 1978 1983 1988

Oper Expend/Faculty

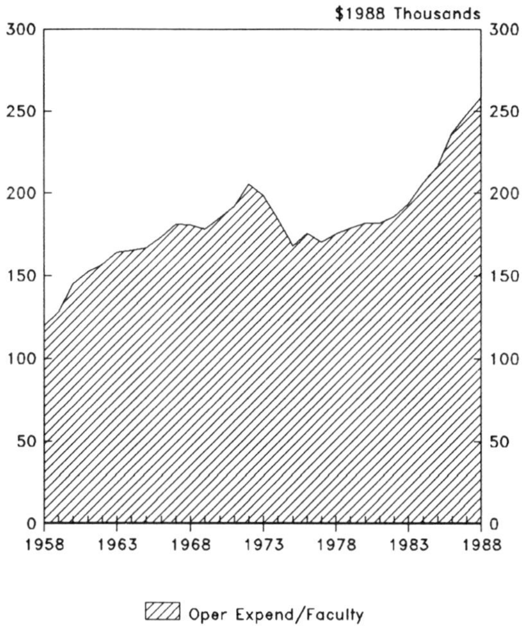

Figure 2-55: Doctoral Institution Education Expenditures per Student

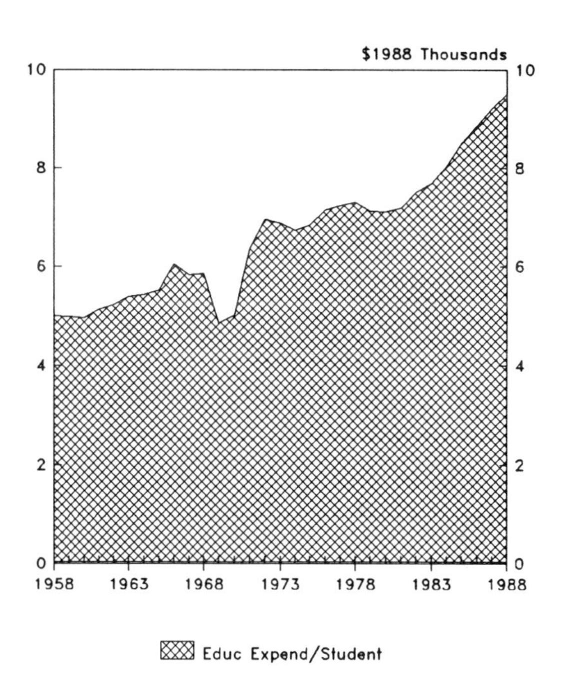

NOTE: Financial data are expressed in 1988 constant dollars to reflect real long-term growth trends.

DEFINITION OF TERMS: Doctoral institutions are higher education institutions that have granted an average of 10 or more Ph.D. degrees per year in the natural sciences or engineering over the past two decades; they include 116 public and 69 private institutions. *Operating Expenditures* consist of educational and general current-fund expenditures for instruction, research, public service, academic support, student services, institutional support, operation and maintenance of plant, scholarships and fellowships, and educational and mandatory transfers for debt service, and for auxiliary enterprises and federally funded research and development centers. They exclude expenditures from institutional plant-fund accounts and Pell Grants. *Educational Expenditures* includes instructional expenditures, including departmental research not separately budgeted; current operating expenditures for libraries, operation and maintenance of plant, scholarships and fellowships, and student services. *Faculty* members include all instructional members of the instruction or research staff whose major regular assignment is instruction, including those with release time for research. *Students* include all full-time students plus a full-time equivalent of part-time students as reported by doctoral institutions.

SOURCE: National Science Foundation, Division of Policy Research and Analysis. Database: CASPAR. Some of the data within this database are estimates, incorporated where there are discontinuities within data series or gaps in data collection. Primary data sources: U.S. Department of Education, National Center for Education Statistics, Higher Education General Information Survey (HEGIS): Salaries, Tenure, and Fringe Benefits of Full-time Instructional Faculty; Fall Enrollment in Institutions of Higher Education; Financial Statistics of Institutions of Higher Education; American Council on Education; National Association of State Universities and Land Grant Colleges.

Doctoral Institution Operating Revenues: Sources

Total doctoral institution operating revenues reached nearly $66 billion in 1988. State funds, income from tuition, and funds from hospitals and auxiliary enterprises contributed most to this growth. Since 1973, federal funds for R&D have accounted for a steady 10 percent share of total doctoral revenues, with federally financed research and development centers accounting for around 7 percent.

Figure 2-56: Revenues of Doctoral Institutions by Source of Funds

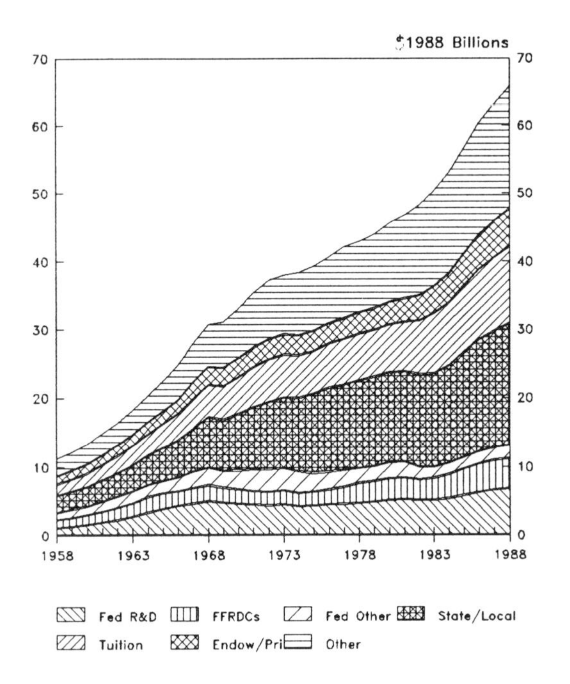

Figure 2-57: Distribution of Revenues of Doctoral Institutions by Source of Funds

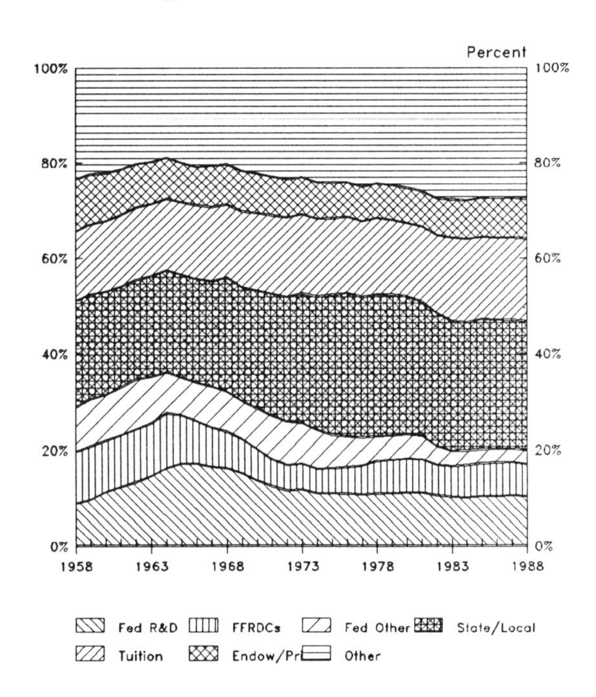

NOTE: Data series within the figures are not overlapped; top line represents total. Financial data are expressed in 1988 constant dollars to reflect real long-term growth trends.

DEFINITION OF TERMS: Doctoral institutions are higher education institutions that have granted an average of 10 or more Ph.D. degrees per year in the natural sciences or engineering over the past two decades; they include 116 public and 69 private institutions. *Federal R&D* includes grants and contracts for specific research projects. *FFRDCs* includes funds for university-administered federally funded research and development centers. *Other Federal* includes dollars appropriated or made available by the federal government to public or private institutions of higher education for current operating expenses, such as land-grant appropriations and revenue sharing funds, or other types of programs such as administrative allowances for student aid; excludes Pell Grants. *Tuition* includes all student assessments for current operating purposes. *State/Local* sources include dollars appropriated or made available by state and local governments to public or private institutions of higher education for current operating expenses and or for specific projects or programs. *Endowment/Private* income includes the unrestricted income of endowment and similar funds; income from private gifts and grants that are directly related to instruction, research, or public service. *Other* includes sales and services of educational activities and revenues derived from the sales of goods or services that are incidental to the conduct of instruction, research, or public service, including hospitals.

SOURCE: National Science Foundation, Division of Policy Research and Analysis. Database: CASPAR. Some of the data within this database are estimates, incorporated where there are discontinuities within data series or gaps in data collection. Primary data source: U.S. Department of Education, National Center for Education Statistics, Higher Education General Information Survey (HEGIS): Financial Statistics of Institutions of Higher Education.

Doctoral Institution Operating Revenues: By Governance

During the 1960s and early 1970s, public doctoral institution total revenues grew faster than those of private institutions, raising the public-institution share of total doctoral revenues from half in 1958 to two-thirds in 1973, where it has remained steady. In 1988, public doctoral institutions received $44 billion; private doctoral institutions received $22 billion.

Figure 2-58: Operating Revenues of Doctoral Institutions by Institution Governance

Figure 2-59: Distribution of Operating Revenues of Doctoral Institutions by Institution Governance

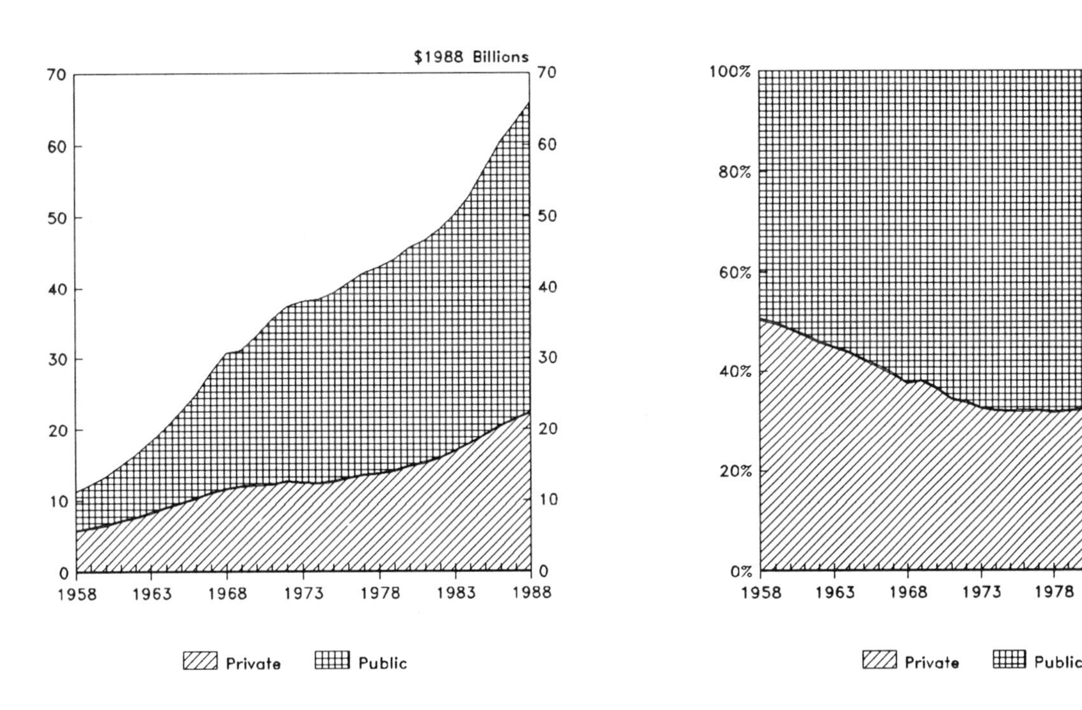

NOTE: Data series within the figures are not overlapped; top line represents total. Financial data are expressed in 1988 constant dollars to reflect real long-term growth trends.

DEFINITION OF TERMS: Operating revenues consist of educational and general current-fund revenues from federal, state, and local appropriations (excluding Pell Grants); tuition income; government grants and contracts; private gifts, grants, and endowment income; sales and services of educational activities; and other revenues, including hospitals and FFRDCs. *Private* doctoral institutions are higher education institutions that have granted an average of 10 or more Ph.D. degrees per year in the natural sciences or engineering over the past two decades, and are under the control of--or affiliated with--non-profit, independent organizations with or without religious affiliation; they include 69 institutions. *Public* doctoral institutions are higher education institutions that have granted an average of 10 or more Ph.D. degrees per year in the natural sciences or engineering over the past two decades, and are under the control of--or affiliated with--federal, state, local, state and local, or state-related agencies; they include 116 institutions.

SOURCE: National Science Foundation, Division of Policy Research and Analysis. Database: CASPAR. Some of the data within this database are estimates, incorporated where there are discontinuities within data series or gaps in data collection. Primary data source: U.S. Department of Education, National Center for Education Statistics, Higher Education General Information Survey (HEGIS): Financial Statistics of Institutions of Higher Education.

Operating Revenues: Private Doctoral Institutions

For the past three decades, student tuition and fees have constituted a steadily growing share of private doctoral institution revenues, increasing from 18 percent in 1958 to 26 percent in 1988. The federal share nearly doubled during the 1960s, from 17 percent in 1958 to over 32 percent in 1966, then declined steadily for the past two decades, down to 16 percent in 1988. The shares of private contributions and endowment income have been relatively stable at 9 percent and 6 percent, respectively. The share from other revenues sources--hospitals, auxiliary enterprises and federally funded research and development centers (FFRDCs)--declined from 39 percent in 1958 to 30 percent in 1968, then returned to a 40 percent by 1988.

Figure 2-60: Private Doctoral Institution Operating Revenues by Source

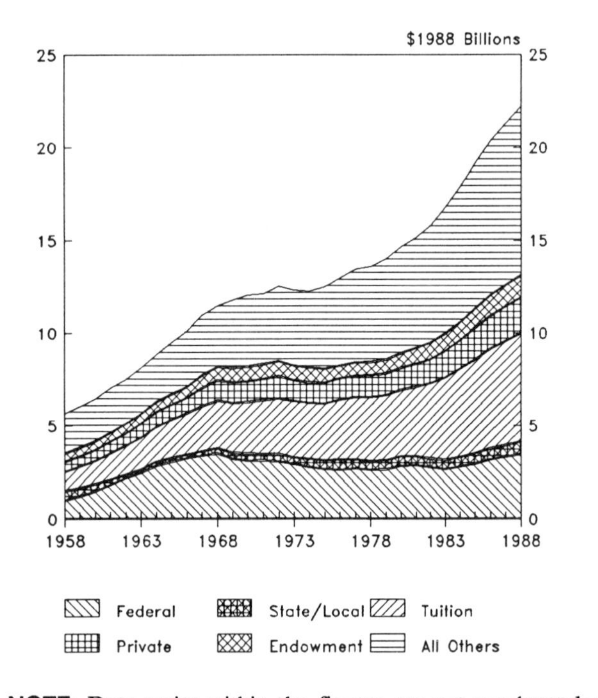

Figure 2-61: Distribution of Private Doctoral Institution Operating Revenues by Source

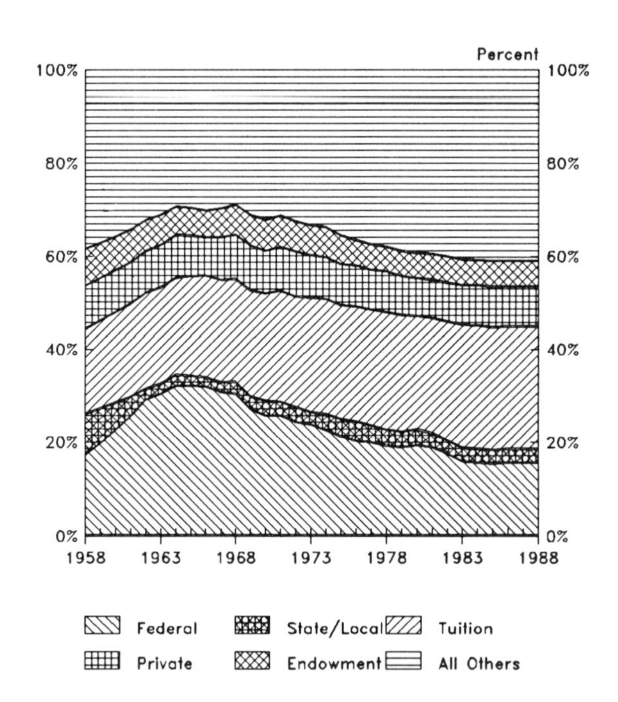

NOTE: Data series within the figures are not overlapped; top line represents total. Financial data are expressed in 1988 constant dollars to reflect real long-term growth trends.

DEFINITION OF TERMS: Private doctoral institutions are higher education institutions that have granted an average of 10 or more Ph.D. degrees per year in the natural sciences or engineering over the past two decades, and are under the control of--or affiliated with--nonprofit, independent organizations with or without religious affiliation; they include 69 institutions.
Federal sources include (1) dollars appropriated or made available by the federal government to public or private institutions of higher education for current operating expenses, such as land-grant appropriations and revenue sharing funds and (2) federal government grants and contracts for specific research projects or other types of programs such as administrative allowances for student aid; excludes Pell Grants. *Tuition* and fees include all assessments against students for current operating purposes. *State/Local* sources include dollars appropriated or made available by state and local governments to public or private institutions of higher education for current operating expenses and or for specific projects or programs. *Private* income includes gifts and grants that are directly related to instruction, research, or public service. *Endowment* income includes the unrestricted income of endowment and similar funds. *Other* includes sales and services of educational activities, and auxiliary enterprises including hospitals and FFRDCs.

SOURCE: National Science Foundation, Division of Policy Research and Analysis. Database: CASPAR. Some of the data within this database are estimates, incorporated where there are discontinuities within data series or gaps in data collection. Primary data source: U.S. Department of Education, National Center for Education Statistics, Higher Education General Information Survey (HEGIS): Financial Statistics of Institutions of Higher Education.

Operating Revenues: Public Doctoral Institutions

For the past three decades, state and local funds have constituted a large and stable share of public doctoral institution revenues, with a 39 percent share during the late-1988s. The federal government share has declined steadily for the past two decades, from 20 percent in 1968 to 12 percent in 1988. The revenue share from tuition has been stable at around 11 percent over the three-decade period; private contributions and endowment income together have averaged about 5 percent. Other revenues sources--hospitals, auxiliary enterprises and federally funded research and development centers (FFRDCs)--have maintained a relatively steady 30-percent share.

Figure 2-62: Public Doctoral Institution Operating Revenues by Source

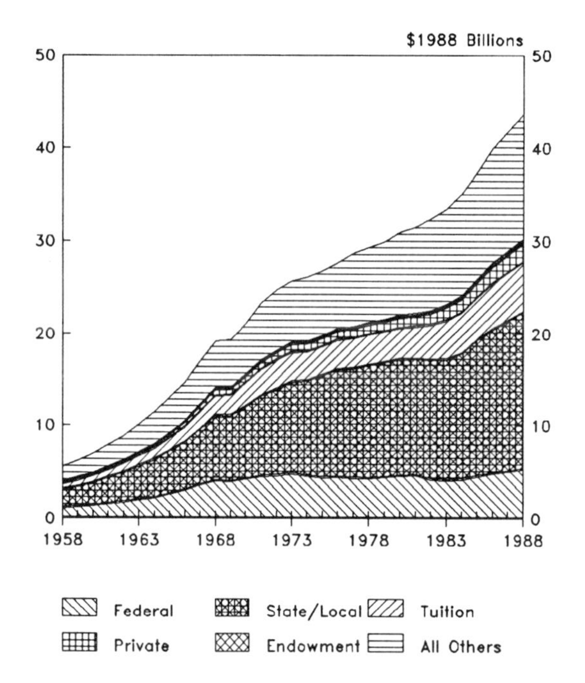

Figure 2-63: Distribution of Public Doctoral Institution Operating Revenues by Source

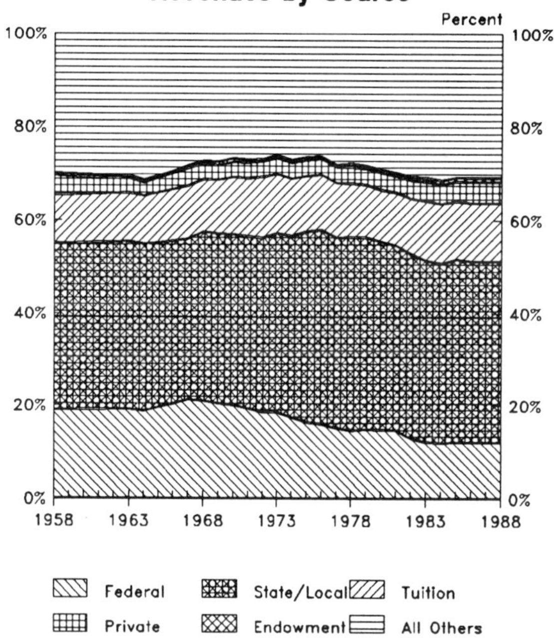

NOTE: Data series within the figures are not overlapped; top line represents total. Financial data are expressed in 1988 constant dollars to reflect real long-term growth trends.

DEFINITION OF TERMS: Public doctoral institutions include those institutions (1) which have granted an average of ten or more Ph.D. degrees per year in the natural sciences or engineering over the past two decades and (2) are under the control of--or affiliated with--federal, state, local, state and local, or state-related agencies. They include 116 institutions. *Federal* sources include (1) dollars appropriated or made available by the federal government to public or private institutions of higher education for current operating expenses, such as land-grant appropriations and revenue sharing funds and (2) federal government grants and contracts for specific research projects or other types of programs such as administrative allowances for student aid. Excludes Pell Grants. *Tuition* and fees include all student assessments for current operating purposes. *State/Local* sources include dollars appropriated or made available by state and local governments to public or private institutions of higher education for current operating expenses and or for specific projects or programs. *Private* income includes private gifts and grants that are directly related to instruction, research, or public service. *Endowment income* includes the unrestricted income of endowment and similar funds. *Other* includes sales and services of educational activities and auxillary enterprises including hospitals and FFRDCs.

SOURCE: National Science Foundation, Division of Policy Research and Analysis. Database: CASPAR. Some of the data within this database are estimates, incorporated where there are discontinuities within data series or gaps in data collection. Primary data source: U.S. Department of Education, National Center for Education Statistics, Higher Education General Information Survey (HEGIS): Financial Statistics of Institutions of Higher Education.

ACADEMIC PERSONNEL

Total Academic Faculty

With rapid increases in student enrollments during the 1960s and 1970s, the total number of academic faculty increased from 270,000 in 1958 to a peak of 750,000 in 1983, then declined to 720,000 by 1988. For the past three decades, the distribution of faculty among types of institutions has remained nearly constant, with 35 percent in doctoral institutions, 40 percent in comprehensive institutions, and 25 percent in 2-year institutions.

Figure 2-64: Academic Faculty by Institution Type

Thousands
800
600
400
200
0
1958
1963
1968
1973
1978
1983
1988
Doctoral
Comprehensive
Two-Year

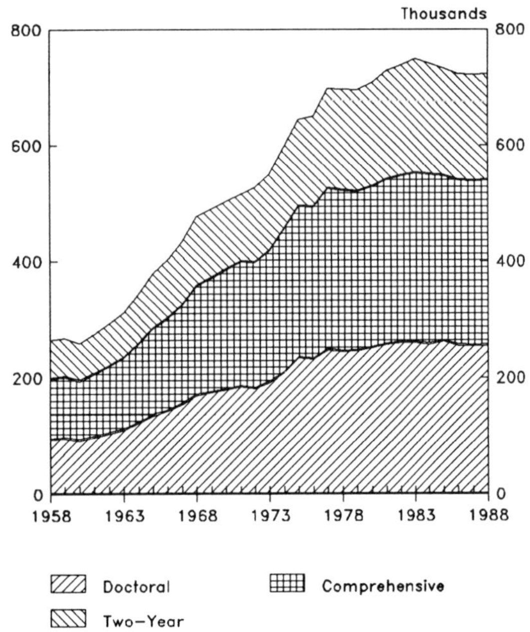

Figure 2-65: Distribution of Academic Faculty by Institution Type

Percent
100%
80%
60%
40%
20%
0%
1958
1963
1968
1973
1978
1983
1988
Doctoral
Comprehensive
Two-Year

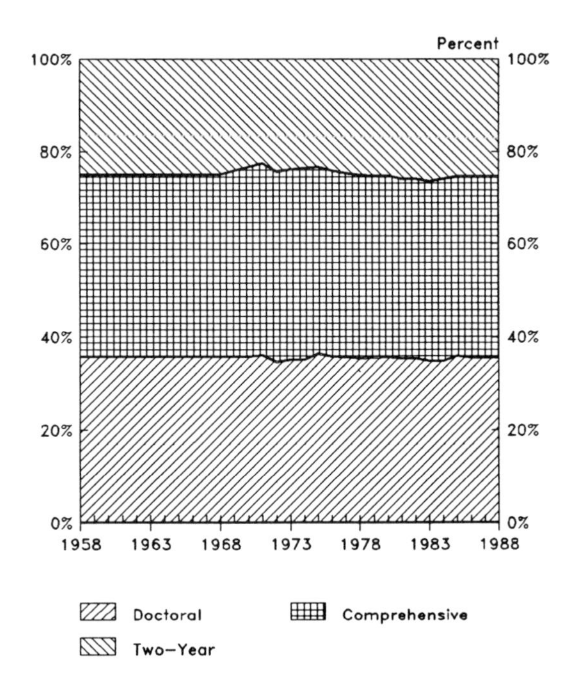

NOTE: Data series within the figures are not overlapped; top line represents total.

DEFINITION OF TERMS: Faculty include all instructional members of the instruction or research staff whose major regular assignment is instruction, including those with release time for research. *Doctoral* institutions are higher education institutions that have granted an average of 10 or more Ph.D. degrees per year in the natural sciences or engineering over the past two decades; they include 116 public and 69 private institutions. *Comprehensive* institutions are those that grant at least half of their degrees for courses of study that normally require 4 or more years to complete; they include 370 public and 854 private institutions. *Two-year* institutions award primarily 2-year associate or technician degrees; they include 902 public and 486 private institutions.

SOURCE: National Science Foundation, Division of Policy Research and Analysis. Database: CASPAR. Some of the data within this database are estimates, incorporated where there are discontinuities within data series or gaps in data collection. Primary data source: American Council on Education; U.S. Department of Education, National Center for Education Statistics; National Association of State Universities and Land Grant Colleges.

Academic Scientists and Engineers

Scientists and engineers employed by universities and colleges in faculty and non-faculty positions have increased steadily from 120,000 in 1958 to 330,000 in 1988 (full-time equivalent). For the past three decades, doctoral institutions have consistently employed 60 percent of all academic scientists and engineers.

Figure 2-66: Academic Scientists and Engineers (FTE) by Institution Type and Governance

Figure 2-67: Distribution of Academic Scientists and Engineers (FTE) by Institution Type and Governance

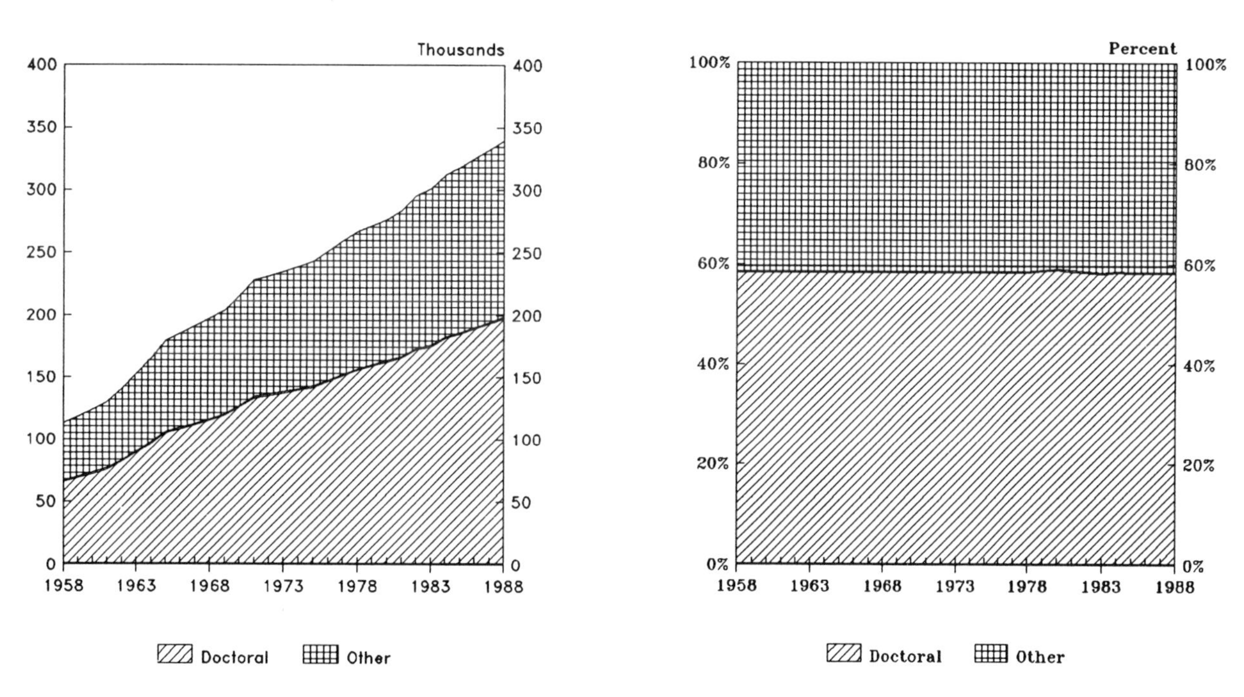

NOTE: Data series within the figures are not overlapped; top line represents total.

DEFINITION OF TERMS: Scientists and engineers (FTE) include all professional employees--faculty and non-faculty personnel and post-doctorates--employed full-time by higher education institutions, plus a full-time equivalent for part-time employees, within the broad fields of physical sciences, engineering, environmental sciences, life and health sciences, mathematics and computer sciences, and social and behavioral sciences. *Doctoral* institutions are higher education institutions that have granted an average of 10 or more Ph.D. degrees per year in the natural sciences or engineering over the past two decades; they include 116 public and 69 private institutions. *Other* institutions include 1,124 comprehensive institutions that grant at least half of their degrees for courses of study that normally require 4 or more years to complete, and 1,388 2-year institutions that primarily award 2-year associate or technician degrees.

SOURCE: National Science Foundation, Division of Policy Research and Analysis. Database: CASPAR. Some of the data within this database are estimates, incorporated where there are discontinuities within data series or gaps in data collection. Primary data source: National Science Foundation, Division of Science Resource Studies, Survey of Scientific and Engineering Personnel Employed at Universities and Colleges.

Doctoral Institution Faculty

Doctoral institutions employed 255,000 faculty members in 1988, roughly stable since the mid-1970s. Over the past three decades, the public doctoral institution share of faculty members has slowly increased from 70 percent in 1958 to over 75 percent in 1988.

Figure 2-68: Doctoral Institution Faculty by Institution Governance

Figure 2-69: Distribution of Doctoral Institution Faculty by Institution Governance

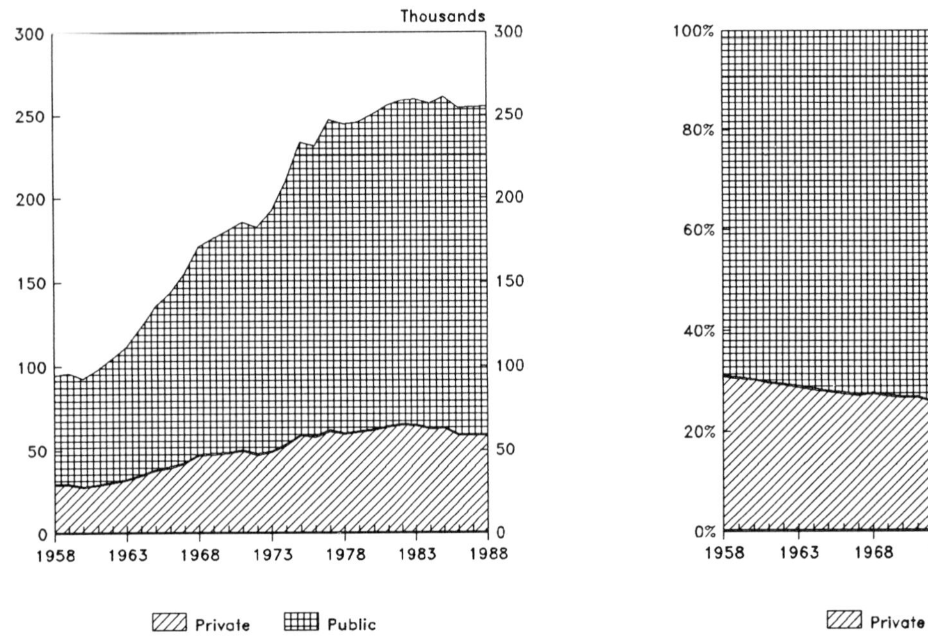

NOTE: Data series within the figures are not overlapped; top line represents total.

DEFINITION OF TERMS: Faculty include all instructional members of the instruction or research staff whose major regular assignment is instruction, including those with release time for research. *Private* doctoral institutions are institutions that have granted an average of 10 or more Ph.D. degrees per year in the natural sciences or engineering over the past two decades, and are under the control of--or affiliated with--non-profit, independent organizations with or without religious affiliation; they include 69 institutions. *Public* doctoral institutions are institutions that have granted an average of 10 or more Ph.D. degrees per year in the natural sciences or engineering over the past two decades, and are under the control of--or affiliated with--federal, state, local, state and local, or state-related agencies; they include 116 institutions.

SOURCE: National Science Foundation, Division of Policy Research and Analysis. Database: CASPAR. Some of the data within this database are estimates, incorporated where there are discontinuities within data series or gaps in data collection. Primary data source: U.S. Department of Education, National Center for Education Statistics, Higher Education General Information Survey (HEGIS): Salaries, Tenure, and Fringe Benefits of Full-time Instructional Faculty; American Council on Education; National Association of Universities and Land Grant Colleges.

Doctoral Institution Faculty: Per Student and Degree

In the late 1970s, the student-to-faculty ratio within doctoral institutions returned to 1950s levels, where it remained stable throughout the 1980s. During the 1950s, the growth rate in student enrollments exceeded the growth rate in numbers of faculty, raising the over-all student-to-faculty ratio. While enrollment growth slowed in the early 1970s, doctoral institutions, as a whole, continued to employ additional faculty. By 1978, the student-to-faculty ratio returned to the 1958 level. The ratio of total degrees awarded per faculty member followed a similar pattern and has remained stable for the past decade.

Figure 2-70: Student-to-Faculty Ratio in Doctoral Institutions

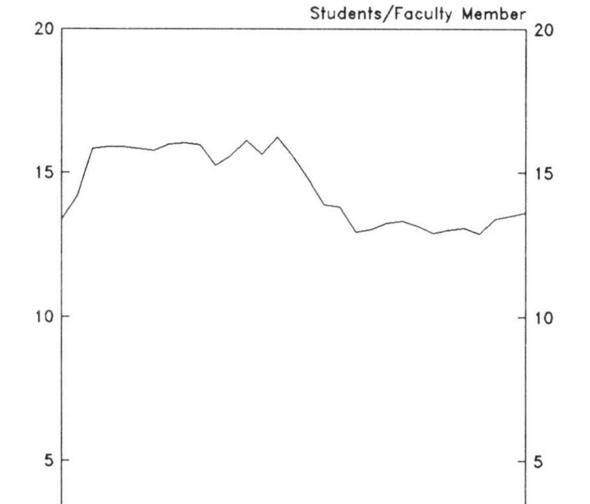

Figure 2-71: Degrees Awarded-per-Faculty Ratio in Doctoral Institutions

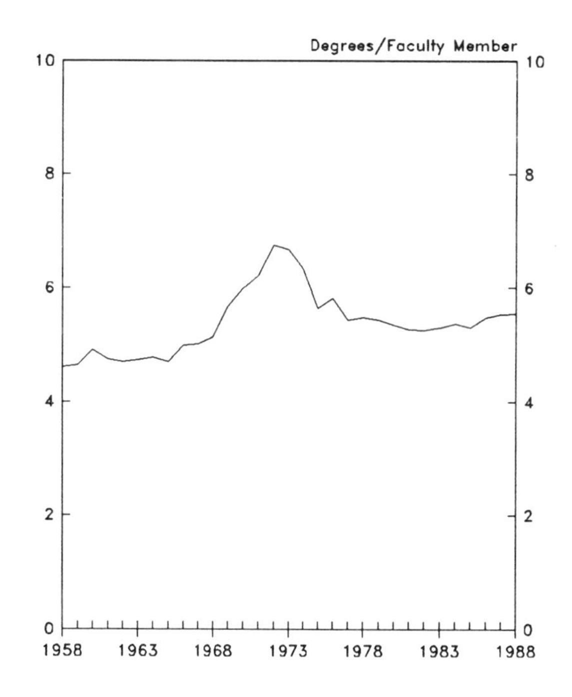

DEFINITION OF TERMS: *Student-to-Faculty Ratio* is derived for each year by dividing total number of students (FTE) by total number of faculty. *Degrees Awarded-per-Faculty Ratio* is derived for each year by dividing the total number of degrees awarded by doctoral institutions by the total number of faculty. Students (FTE) include all full-time students plus a full-time-equivalent of part-time students as reported by doctoral institutions. Degrees include all degrees awarded by doctoral institutions in all academic disciplines, both undergraduate and graduate. Faculty include all instructional members of the instruction or research staff of doctoral institutions whose major regular assignment is instruction, including those with release time for research. Doctoral institutions are higher education institutions that have granted an average of 10 or more Ph.D. degrees per year in the natural sciences or engineering over the past two decades; they include 116 public and 69 private institutions.

SOURCE: National Science Foundation, Division of Policy Research and Analysis. Database: CASPAR. Some of the data within this data base are estimates, incorporated where there are discontinuities within data series or gaps in data collection. Primary data source: U.S. Department of Education, National Center for Education Statistics, Higher Education General Information Survey (HEGIS): Fall Enrollment in Institutions of Higher Education; Degrees and Other Formal Awards Conferred, Salaries, Tenure, and Fringe Benefits of Full-time Instructional Faculty; American Council on Education; National Association of State Universities and Land Grant Colleges.

Doctoral Institution Scientists and Engineers

The number of scientists and engineers (FTE) employed by doctoral institutions, in both faculty and non-faculty positions, has increased steadily from 66,000 in 1958 to nearly 200,000 in 1988. For the past two decades, public doctoral institutions have employed nearly 70 percent of all doctoral institution scientists and engineers.

Figure 2-72: Scientists and Engineers (FTE) in Doctoral Institutions by Institution Governance

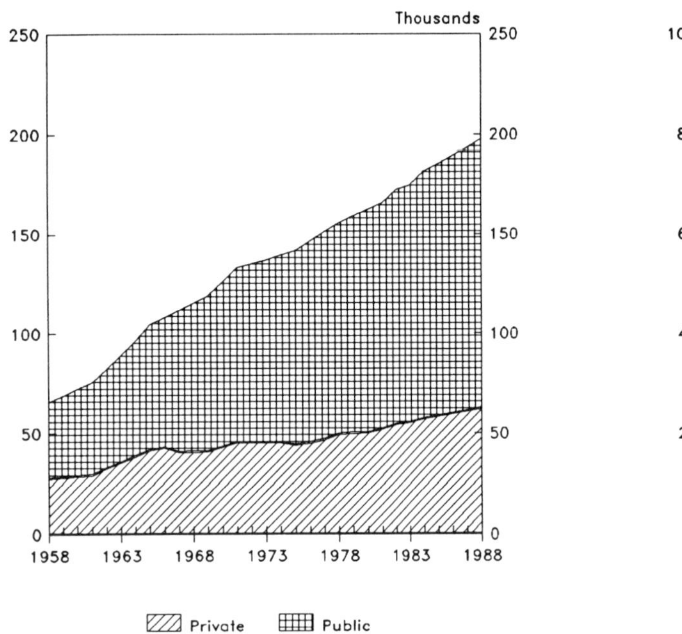

Figure 2-73: Distribution of Scientists and Engineers (FTE) in Doctoral Institutions by Institution Governance

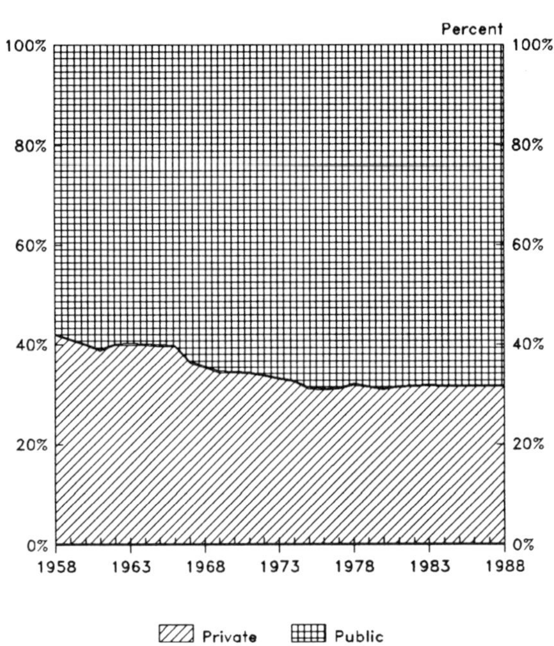

NOTE: Data series within the figures are not overlapped; top line represents total.

DEFINITION OF TERMS: Scientists and engineers include all professional employees--faculty, non-faculty, and post-doctorate personnel--employed by higher education institutions (plus a full-time equivalent for part-time employees), within the broad fields of physical sciences, engineering, environmental sciences, life and health sciences, mathematics and computer sciences, and social and behavioral sciences. *Private* doctoral institutions are institutions that have granted an average of 10 or more Ph.D. degrees per year in the natural sciences or engineering over the past two decades, and are under the control of--or affiliated with--non-profit, independent organizations with or without religious affiliation; they include 69 institutions. *Public* doctoral institutions are institutions that have granted an average of 10 or more Ph.D. degrees per year in the natural sciences or engineering over the past two decades, and are under the control of--or affiliated with--federal, state, local, state and local, or state-related agencies; they include 116 institutions.

SOURCE: National Science Foundation, Division of Policy Research and Analysis. Database: CASPAR. Some of the data within this database are estimates, incorporated where there are discontinuities within data series or gaps in data collection. Primary data source: National Science Foundation, Division of Science Resource Studies, Survey of Scientific and Engineering Personnel Employed at Universities and Colleges.

Doctoral Institution Research Personnel

The number of investigators (FTE)--faculty and non-faculty--has increased from 25,000 in 1958 to 63,000 by 1988. The public doctoral institutions share of investigators (FTE) rose from 50 percent in 1958 to 65 percent in 1988.

Figure 2-74: Investigators (FTE) in Doctoral Institutions by Institution Governance

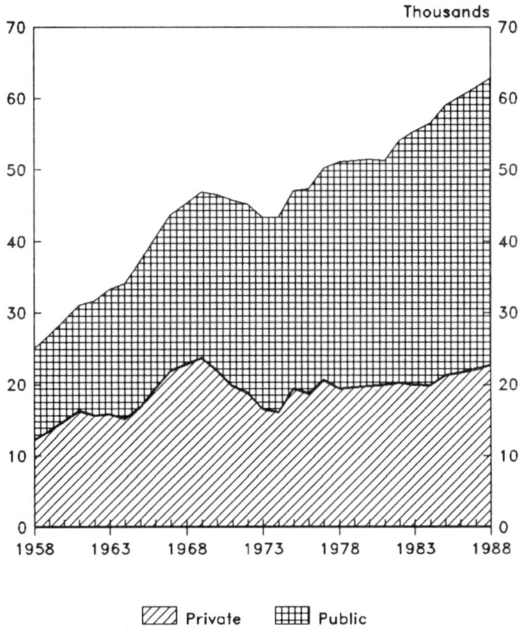

Figure 2-75: Distribution of Investigators (FTE) in Doctoral Institutions by Institution Governance

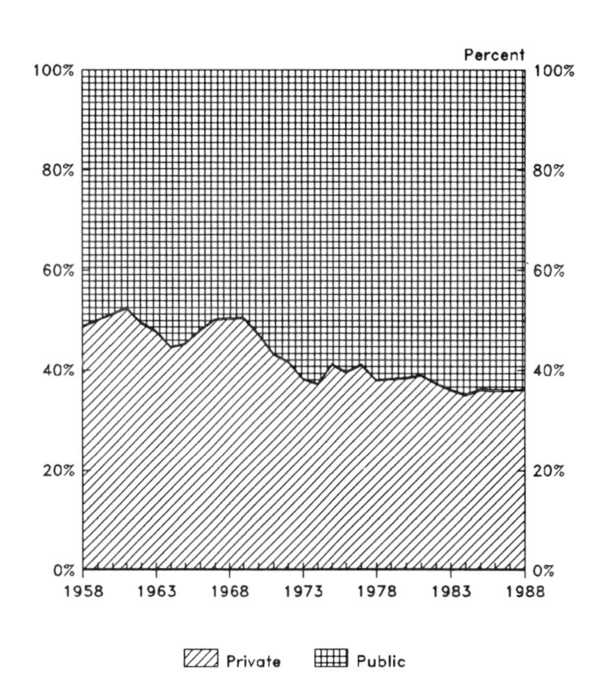

NOTE: Data series within the figures are not overlapped; top line represents total.

DEFINITION OF TERMS: Investigators (FTE) include scientists and engineers (in the physical sciences, engineering, environmental sciences, life and health sciences, mathematics and computer sciences, and social and behavioral sciences) conducting separately budgeted academic R&D; their numbers are estimated by the fraction of faculty time spent in research activities, non-faculty scientists and engineers employed to conduct research in campus facilities (except FFRDCs), post-doctoral researchers working in academic institutions. *Private* doctoral institutions are higher education institutions that have granted an average of 10 or more Ph.D. degrees per year in the natural sciences or engineering over the past two decades, and are under the control of--or affiliated with--non-profit, independent organizations with or without religious affiliation; they include 69 institutions. *Public* doctoral institutions are higher education institutions that have granted an average of 10 or more Ph.D. degrees per year in the natural sciences or engineering over the past two decades, and are under the control of--or affiliated with--federal, state, local, state and local, or state-related agencies; they include 116 institutions.

SOURCE: National Science Foundation, Division of Policy Research and Analysis. Database: CASPAR. Some of the data within this database are estimates, incorporated where there are discontinuities within data series or gaps in data collection. Primary data source: National Science Foundation, Division of Science Resource Studies, Survey of Scientific and Engineering Personnel Employed at Universities and Colleges.

Doctoral Institution S&E Personnel Ratios

Within doctoral institutions, the over-all ratio of scientists and engineers (FTE) to total faculty (FTE) has slowly increased during the 1980s; for private doctoral institutions, the number of scientists and engineers employed (faculty and non-faculty) exceeds the total number of faculty members in all academic disciplines. The ratio of investigators (FTE) to all scientists and engineers (FTE) has been stable for the past decade.

Figure 2-76: Ratio of FTE Scientists and Engineers to All Faculty in Doctoral Institutions

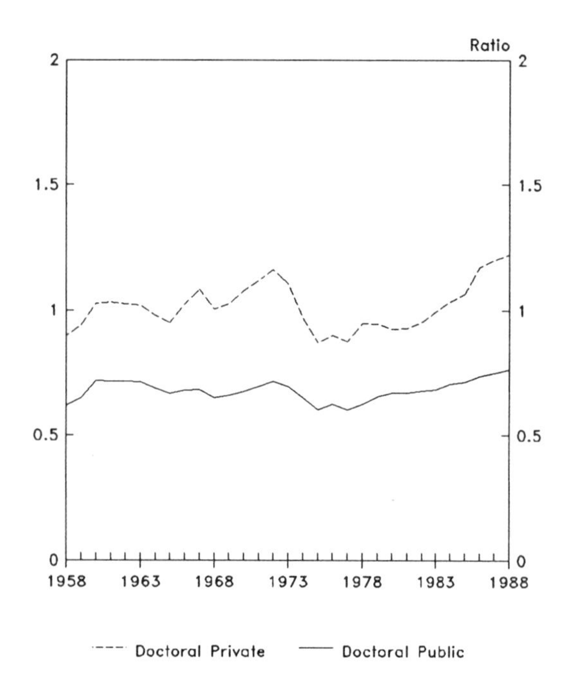

Figure 2-77: Ratio of FTE Investigators to FTE Scientists and Engineers in Doctoral Institutions

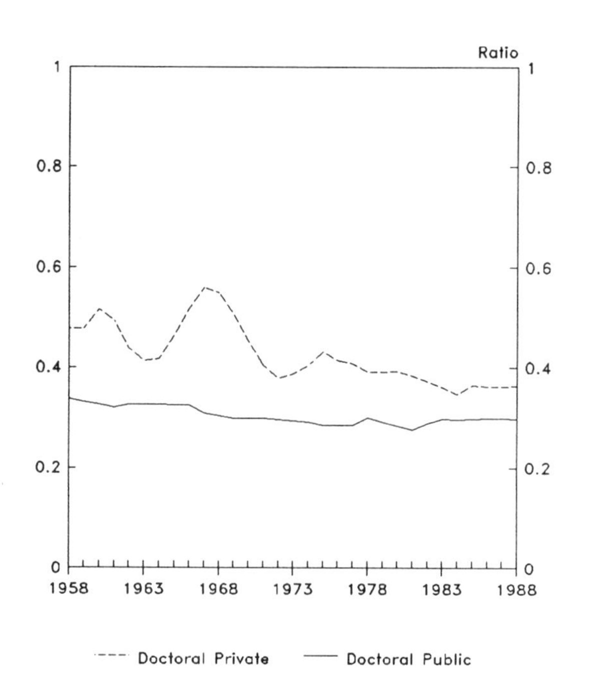

DEFINITION OF TERMS: Faculty include all instructional members of the instruction or research staff whose major regular assignment is instruction, including those with release time for research. FTE scientists and engineers include all professional employees--faculty, non-faculty, and post-doctorate personnel--employed by higher education institutions (plus a full-time equivalent for part-time employees), within the broad fields of physical sciences, engineering, environmental sciences, life and health sciences, mathematics and computer sciences, and social and behavioral sciences. FTE investigators include those scientists and engineers (within the physical sciences, engineering, environmental sciences, life and health sciences, mathematics and computer sciences, and social and behavioral sciences) conducting separately budgeted academic R&D; an estimate derived from the fraction of faculty time spent in research activities, non-faculty scientists and engineers employed to conduct research in campus facilities (except FFRDCs), and post-doctoral researchers working in academic institutions. *Doctoral Public* institutions are higher education institutions that have granted an average of 10 or more Ph.D. degrees per year in the natural sciences or engineering over the past two decades, and are under the control of--or affiliated with--federal, state, local, state and local, or state-related agencies; they include 116 institutions. *Doctoral Private* institutions are higher education institutions that have granted an average of 10 or more Ph.D. degrees per year in the natural sciences or engineering over the past two decades, and are under the control of--or affiliated with--non-profit, independent organizations with or without religious affiliation; they include 69 institutions.

SOURCE: National Science Foundation, Division of Policy Research and Analysis. Database: CASPAR. Some of the data within this database are estimates, incorporated where there are discontinuities within data series or gaps in data collection. Primary data source: National Science Foundation, Division of Science Resource Studies, Survey of Scientific and Engineering Personnel Employed at Universities and Colleges; American Council on Education; National Association of State Universities and Land Grant Colleges.

HIGHER EDUCATION ENROLLMENTS

Higher Education Enrollments: By Institution Type

Total higher-education enrollments rose from 3 million in 1958 to 12.5 million in 1988, with most of the increase occurring by the mid-1970s, primarily in comprehensive and 2-year institutions. Enrollments increased less steeply at doctoral institutions, from 1.3 in 1958 to 3.5 million in 1988, yet have been generally level since the mid-1970s.

Figure 2-78: Enrollment in Academic Institutions by Institution Type and Governance

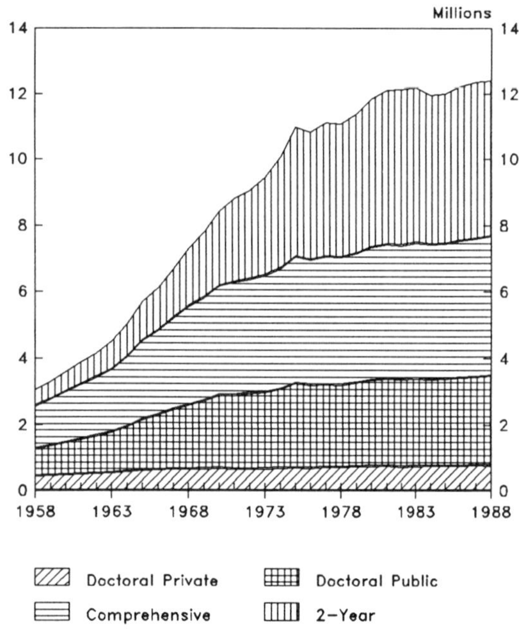

Millions

Doctoral Private
Doctoral Public
Comprehensive
2-Year

Figure 2-79: Distribution of Enrollment in Academic Institutions by Type and Governance

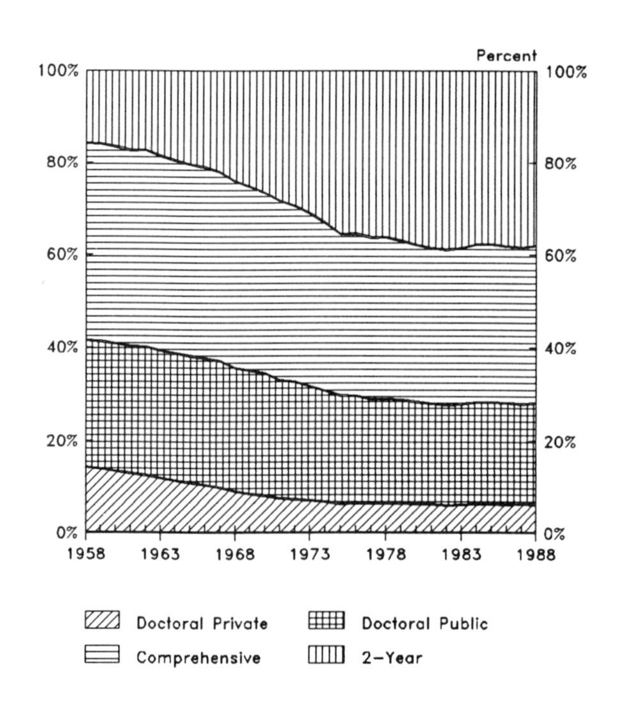

NOTE: Data series within the figures are not overlapped; top line represents total.

DEFINITION OF TERMS: Higher-education enrollments include all full-time students plus a full-time equivalent of part-time students as reported by the institutions. *Private Doctoral* institutions are higher education institutions that have granted an average of 10 or more Ph.D. degrees per year in the natural sciences or engineering over the past two decades, and are under the control of--or affiliated with--non-profit, independent organizations with or without religious affiliation; they include 69 institutions. *Public Doctoral* institutions are higher education institutions that have granted an average of 10 or more Ph.D. degrees per year in the natural sciences or engineering over the past two decades, and are under the control of--or affiliated with--federal, state, local, state and local, or state-related agencies; they include 116 institutions. *Comprehensive* institutions are those that grant at least half of their degrees for courses of study that normally require 4 or more years to complete; they include 854 private and 370 public institutions. *Two-Year* institutions are those that primarily award 2-year associate or technician degrees; they include 486 private and 902 public institutions.

SOURCE: National Science Foundation, Division of Policy Research and Analysis. Database: CASPAR. Some of the data within this database are estimates, incorporated where there are discontinuities within data series or gaps in data collection. Primary data source: U.S. Department of Education, National Center for Education Statistics, Higher Education General Information Survey (HEGIS): Fall Enrollment in Institutions of Higher Education.

Doctoral Institution Enrollments: Undergraduate and Graduate

During the 1960s and 1970s, graduate education assumed a greater role within doctoral institutions. Graduate enrollments, as a share of total enrollments, rose from 20 percent in 1958 to 30 percent by 1976, remaining steady during the 1980s.

Figure 2-80: Undergraduate and Graduate Enrollments in Doctoral Institutions

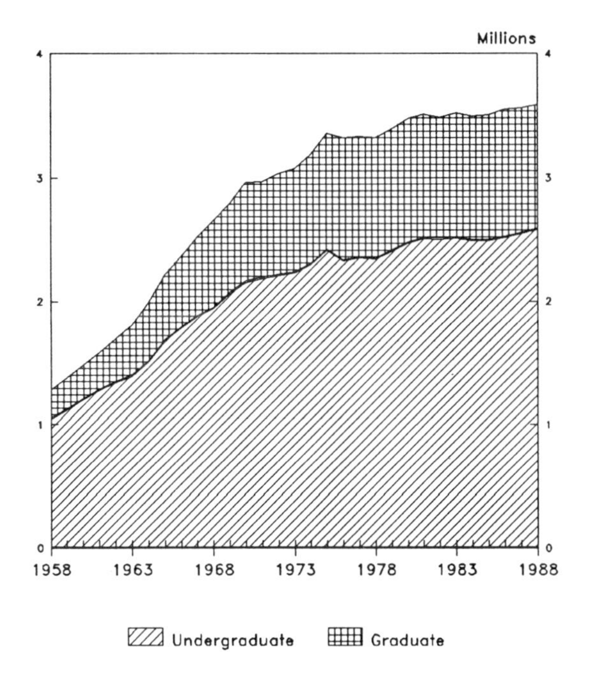

Figure 2-81: Distribution of Undergraduate and Graduate Enrollments in Doctoral Institutions

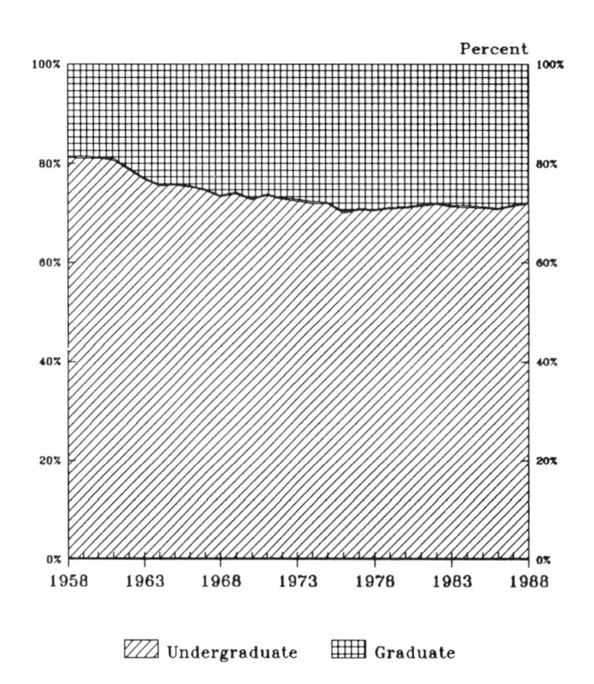

DEFINITION OF TERMS: *Undergraduate* enrollments include all full-time students who are working toward a bachelors or associate degree, or a technician certificate, plus a full-time equivalent of part-time students as reported by institutions; excluded are students of unclassified enrollment status. *Graduate* enrollments include all full-time students (plus a full-time-equivalent of part-time students) who hold the bachelors or equivalent degree, and are working toward an advanced degree, including a first professional degree. Doctoral institutions are higher education institutions that have granted an average of 10 or more Ph.D. degrees per year in the natural sciences or engineering over the past two decades; they include 69 private institutions and 116 public institutions.

SOURCE: National Science Foundation, Division of Policy Research and Analysis. Database: CASPAR. Some of the data within this database are estimates, incorporated where there are discontinuities within data series or gaps in data collection. Primary data source: U.S. Department of Education, National Center for Education Statistics, Higher Education General Information Survey (HEGIS): Fall Enrollment in Institutions of Higher Education.

Doctoral Institution Average Enrollments

Between 1958 and 1988, average enrollments in public doctoral institutions grew to more than twice that of private doctoral institutions.

Figure 2-82: Average Annual Enrollments in Private and Public Doctoral Institutions

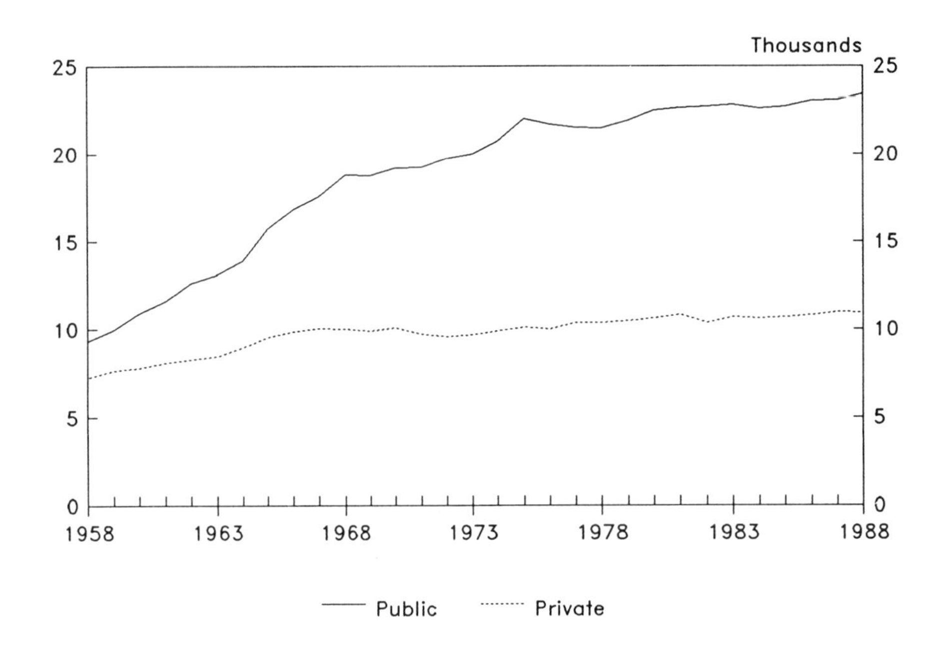

DEFINITION OF TERMS: *Doctoral* institution enrollments include all full-time students plus a full-time equivalent of part-time students as reported by doctoral institutions. *Public* doctoral institutions are institutions that have granted an average of 10 or more Ph.D. degrees per year in the natural sciences or engineering over the past two decades, and are under the control of--or affiliated with--federal, state, local, state and local, or state-related agencies; they include 116 institutions. *Private* doctoral institutions are institutions (1) that have granted an average of 10 or more Ph.D. degrees per year in the natural sciences or engineering over the past two decades and (2) are under the control of--or affiliated with--non-profit, independent organizations with or without religious affiliation; they include 69 institutions.

SOURCE: National Science Foundation, Division of Policy Research and Analysis. Database: CASPAR. Some of the data within this database are estimates, incorporated where there are discontinuities within data series or gaps in data collection. Primary data source: U.S. Department of Education, National Center for Education Statistics, Higher Education General Information Survey (HEGIS): Fall Enrollment in Institutions of Higher Education.

Higher Education Enrollments: Percent Female

Women now comprise over half of all higher-education enrollments. For doctoral institutions, the female share of enrollments grew from 32 percent in 1958 to nearly 50 percent in 1988. In comprehensive and 2-year institutions, the female share of enrollments grew from around 40 percent in 1958 to 55 percent in 1988.

Figure 2-83: Percents of Females Enrolled in Institutions of Higher Education

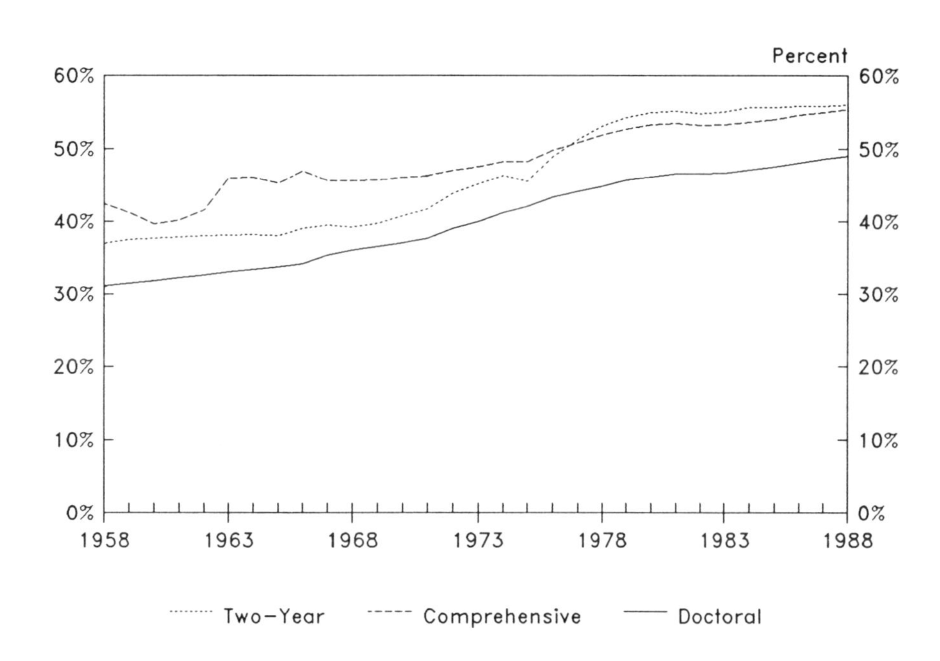

DEFINITION OF TERMS: Higher education enrollments include all full-time students plus a full-time equivalent of part-time students as reported by institutions. *Two-Year* institutions award primarily 2-year associate or technician degrees; they include 902 public and 486 private institutions. *Comprehensive* institutions are those that grant at least half of their degrees for courses of study that normally require 4 or more years to complete; they include 370 public and 854 private institutions. *Doctoral* institutions are institutions that granted an average of 10 or more Ph.D. degrees per year in the natural sciences or engineering over the past two decades; they include 116 public and 69 private institutions.

SOURCE: National Science Foundation, Division of Policy Research and Analysis. Database: CASPAR. Some of the data within this database are estimates, incorporated where there are discontinuities within data series or gaps in data collection. Primary data source: U.S. Department of Education, National Center for Education Statistics, Higher Education General Information Survey (HEGIS): Fall Enrollment in Institutions of Higher Education.

SCIENCE AND ENGINEERING DEGREES

Total S&E Degrees: Type of Degree

During the 1960s and 1970s, the number of science and engineering bachelors degrees awarded annually increased sharply, from 120 thousand in 1958 to 340 thousand in 1974, then more slowly to 377 thousand in the late 1980s. For masters degrees, the number awarded annually increased from 25 thousand in 1958 to 110 thousand in 1988. Likewise, the number of Ph.D. degrees awarded annually increased during the 1960s and early 1970s, rising from 6 thousand in 1958 to 18 thousand in 1974; they have also stabilized in the late 1980s with annual production of about 19 thousand.

Figure 2-84: Degrees Awarded in Science and Engineering by Degree Level

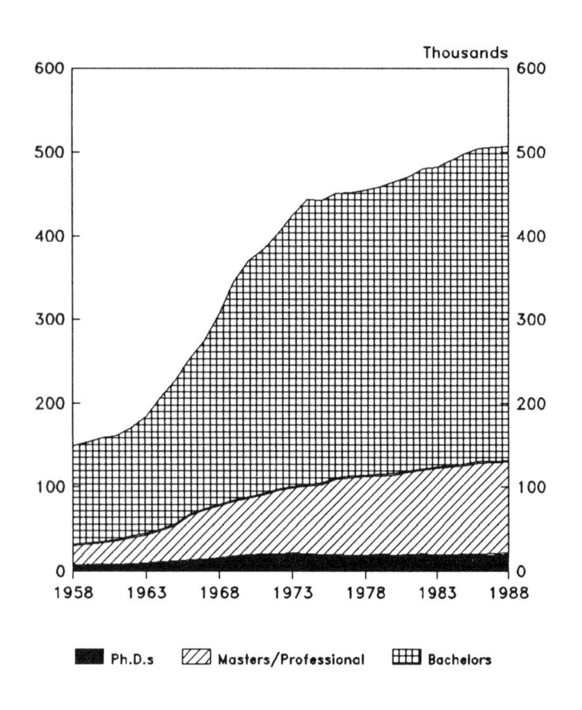

Figure 2-85: Distribution of Degrees Awarded in Science and Engineering by Degree Level

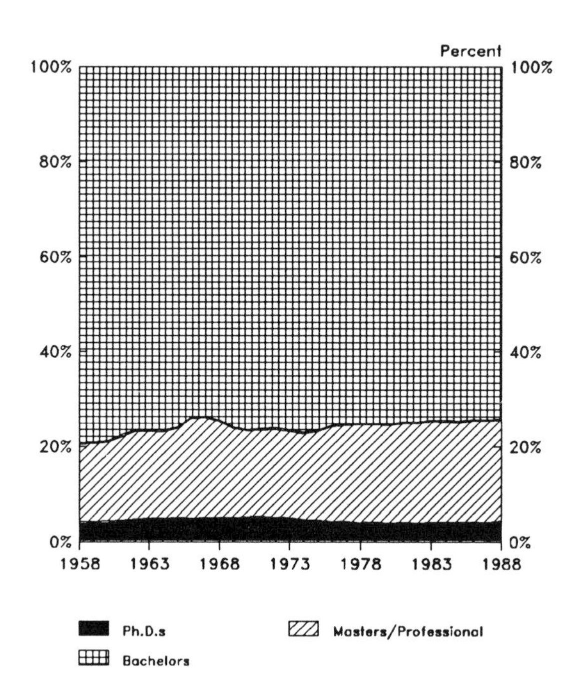

NOTE: Data series within the figures are not overlapped; top line represents total.

DEFINITION OF TERMS: Science and engineering fields are life sciences, including agricultural, biological, medical, and other health sciences; physical sciences including astronomy, chemistry, and physics; engineering including aeronautical and astronautical, chemical, civil, electrical, and mechanical engineering; environmental sciences including oceanography, atmospheric, and earth sciences; mathematics and computer science including all fields of mathematics and computer-related sciences; and social and other behavioral sciences, including economics, political science, psychology, sociology.

SOURCE: National Science Foundation, Division of Policy Research and Analysis. Database: CASPAR. Some of the data within this database are estimates, incorporated where there are discontinuities within data series or gaps in data collection. Primary data source: U.S. Department of Education, National Center for Education Statistics, Higher Education General Information Survey (HEGIS): Degrees and Other Formal Awards Conferred.

Total Bachelors Degrees: S&E and Other Fields

During the 1960s and early 1970s, the number of bachelors degrees awarded annually nearly tripled, from 340 thousand in 1958 to 950 thousand in 1974, then stabilized at 1 million in the late 1980s. The number of science and engineering bachelors degrees increased, from 120 thousand in 1958 to 340 thousand in 1974, and then stabilized at about 375 thousand in the late 1980s. For the past three decades, the share of degrees awarded in the sciences and engineering has remained generally steady, increasing slightly from 34 percent of all bachelors degrees in 1958 to 37 percent in the late-1980s.

Figure 2-86: Bachelors Degrees Awarded in S&E and Other Fields

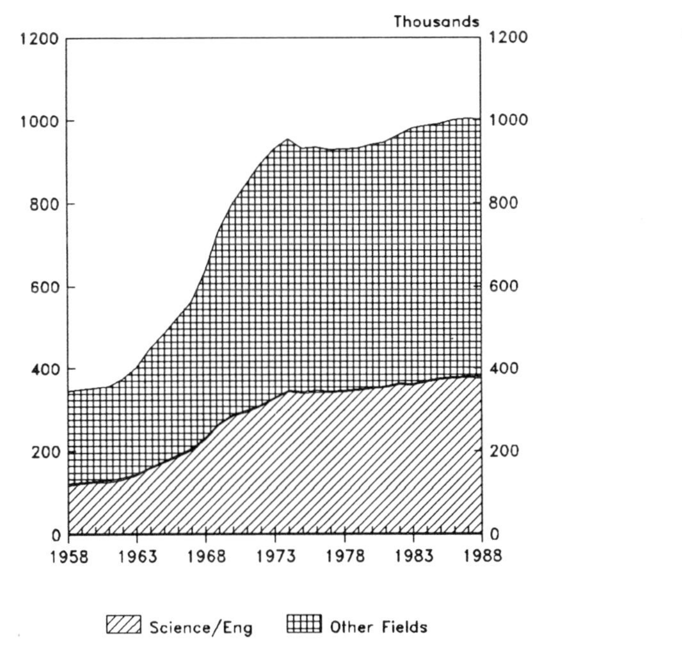

Figure 2-87: Distribution of Bachelors Degrees Awarded in S&E and Other Fields

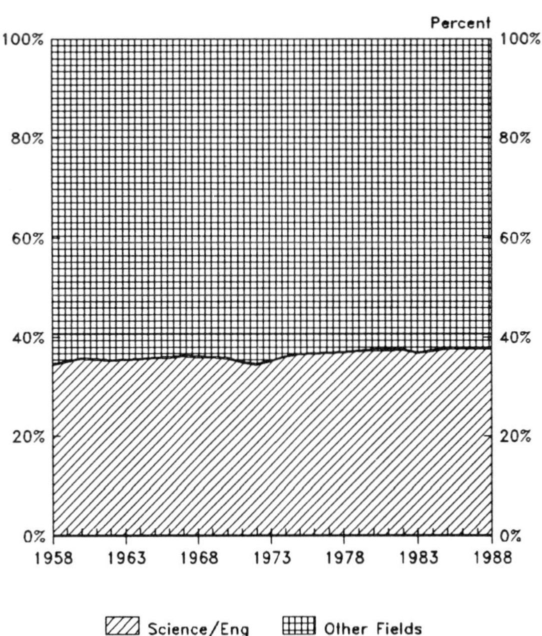

NOTE: Data series within the figures are not overlapped; top line represents total.

DEFINITION OF TERMS: *Science/Engineering* includes bachelors degrees in life sciences, including agricultural, biological, medical, and other health sciences; physical sciences, including astronomy, chemistry, and physics; engineering, including aeronautical and astronautical, chemical, civil, electrical, and mechanical engineering; environmental sciences, including oceanography, atmospheric and earth sciences; mathematics and computer science, including all fields of mathematics and computer-related sciences; and social and other including economics, political science, psychology, sociology. *Other Fields* includes all bachelors degrees other than those awarded in the sciences and engineering.

SOURCE: National Science Foundation, Division of Policy Research and Analysis. Database: CASPAR. Some of the data within this database are estimates, incorporated where there are discontinuities within data series or gaps in data collection. Primary data source: U.S. Department of Education, National Center for Education Statistics, Higher Education General Information Survey (HEGIS): Degrees and Other Formal Awards Conferred.

S&E Bachelors Degrees: Academic Field

During the past three decades, significant shifts have occurred in the proportion of science and engineering bachelors degrees awarded in different academic fields. During the 1960s and early 1970s, degrees in the life and social and behaviorial sciences grew more rapidly than other fields. Between 1978 and 1988, engineering and computer sciences have grown in share of all S&E bachelors degrees awarded. The share within the physical sciences has slowly declined throughout the past three decades.

Figure 2-88: Bachelors Degrees Awarded in S&E by Field of Study

Figure 2-89: Distribution of Bachelors Degrees Awarded in S&E by Field of Study

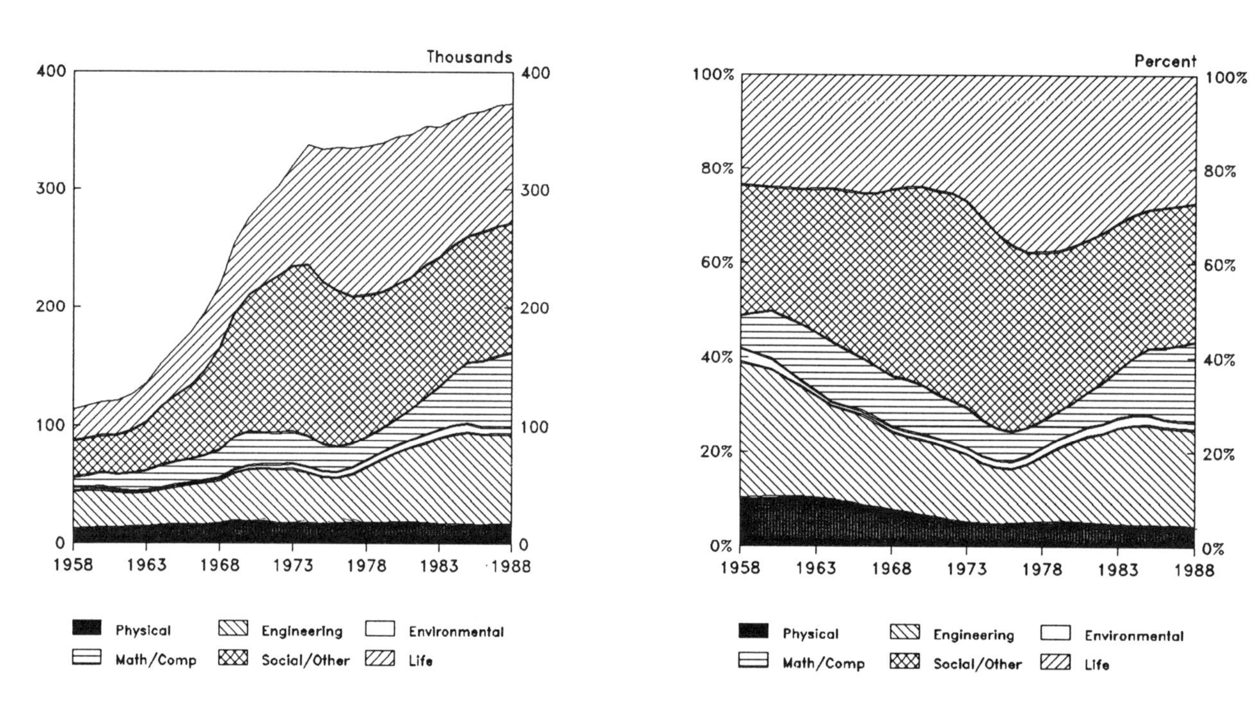

NOTE: Data series within the figures are not overlapped; top line represents total.

DEFINITION OF TERMS: *Physical* includes astronomy, chemistry, and physics. *Engineering* includes aeronautical and astronautical, chemical, civil, electrical, and mechanical engineering. *Environmental* includes oceanography, atmospheric, and earth sciences. *Mathematics/Computer* includes all fields of mathematics and computer-related sciences. *Social/Other* includes economics, political science, psychology, sociology, and public policy-related fields. *Life* includes agricultural, biological, medical, and health sciences.

SOURCE: National Science Foundation, Division of Policy Research and Analysis. Database: CASPAR. Some of the data within this database are estimates, incorporated where there are discontinuities within data series or gaps in data collection. Primary data source: U.S. Department of Education, National Center for Education Statistics, Higher Education General Information Survey (HEGIS): Degrees and Other Formal Awards Conferred.

S&E Bachelors Degrees: Gender

During the past 15 years, the increase in the number of bachelors degrees in the sciences and engineering, although slight, is attributable to additional numbers of women obtaining such degrees. By 1986, the number of S&E bachelor degrees awarded to women rose to nearly 160 thousand, then leveled off in the late-1980s. The number of S&E bachelors degrees awarded annually to men has been generally flat for the past 15 years; fluctuating near the 1980s level of 210 thousand. As a consequence, the share of these degrees awarded to women increased from 20 percent in 1958 to 45 percent by 1980, where it has remained.

Figure 2-90: Bachelors Degrees Awarded in S&E by Gender

Figure 2-91: Distribution of Bachelors Degrees Awarded in S&E by Gender

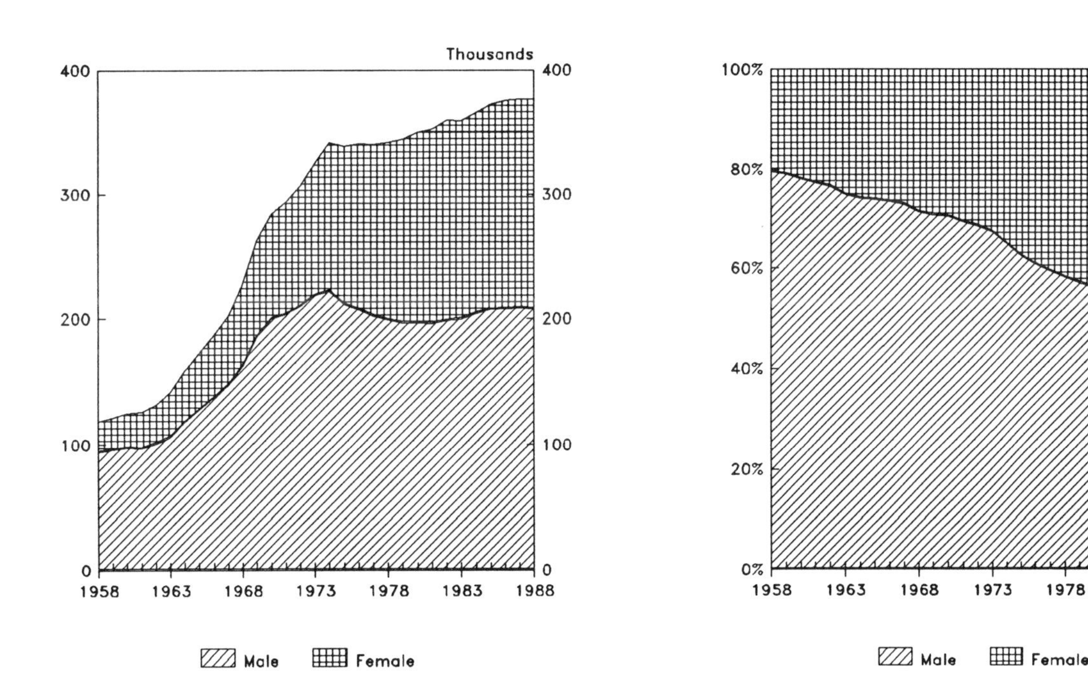

NOTE: Data series within the figures are not overlapped; top line represents total.

DEFINITION OF TERMS: Science and engineering bachelors degrees are awarded in life sciences, including agricultural, biological, medical, and other health sciences; physical sciences including astronomy, chemistry, and physics; engineering including aeronautical and astronautical, chemical, civil, electrical, and mechanical engineering; environmental sciences including oceanography, atmospheric, and earth sciences; mathematics and computer science including all fields of mathematics and computer-related sciences; and social and other, including economics, political science, psychology, sociology.

SOURCE: National Science Foundation, Division of Policy Research and Analysis. Database: CASPAR. Some of the data within this database are estimates, incorporated where there are discontinuities within data series or gaps in data collection. Primary data source: U.S. Department of Education, National Center for Education Statistics, Higher Education General Information Survey (HEGIS): Degrees and Other Formal Awards Conferred.

Total Ph.D. Degrees: S&E and Other Fields

During the 1960s and early 1970s, the total numbers of Ph.D. degrees awarded annually increased steeply, from 9 thousand in 1958 to 35 thousand in 1974. Awards declined to 33 thousand in 1978, then rose to 35 thousand again in 1988. Similarly, the numbers of Ph.D. degrees in science and engineering also fluctuated, from 6 thousand in 1958 to 18 thousand in 1974, down to 17 thousand in 1978, and up to over 20 thousand in 1988. The share of total Ph.D. degrees awarded in the sciences and engineering dipped from 65 percent in 1958 to nearly 50 percent in 1978, before increasing to 57 percent in the late-1980s.

Figure 2-92: Ph.D. Degrees Awarded in S&E and Other Fields

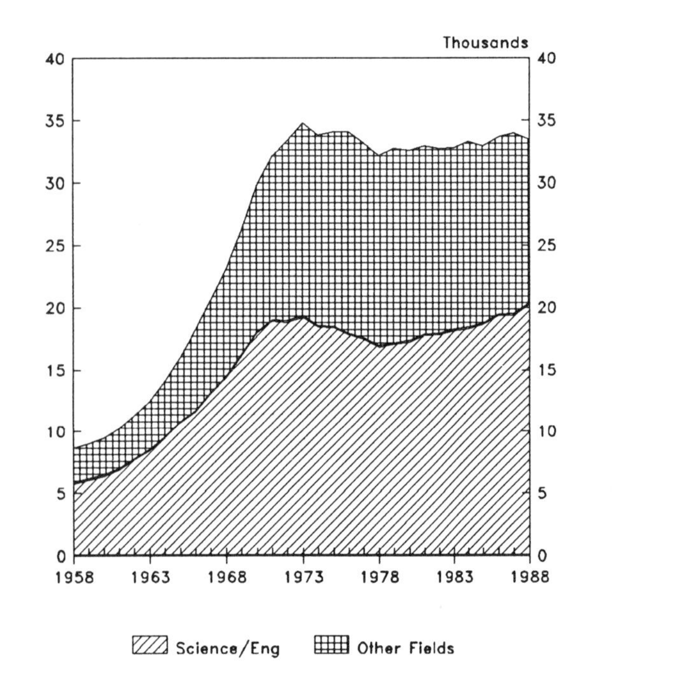

Figure 2-93: Distribution of Ph.D. Degrees Awarded in S&E and Other Fields

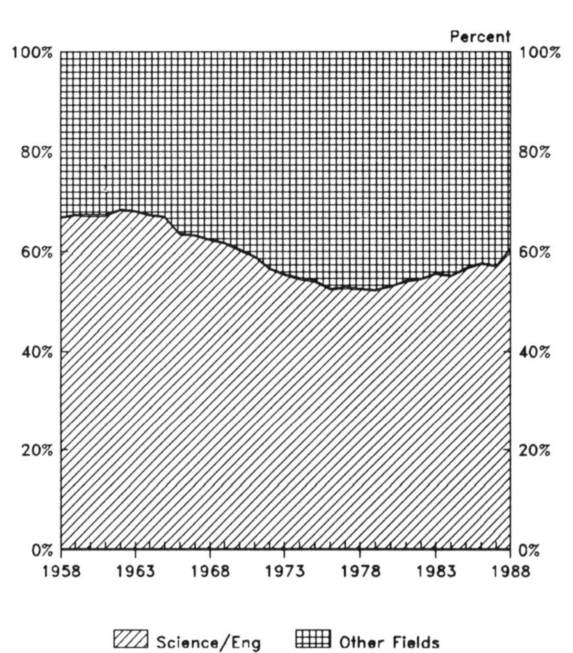

NOTE: Data series within the figures are not overlapped; top line represents total.

DEFINITION OF TERMS: *Science/Engineering* Ph.D. degrees are awarded in life sciences, including agricultural, biological, medical, and other health sciences; physical sciences including astronomy, chemistry, and physics; Engineering including aeronautical and astronautical, chemical, civil, electrical, and mechanical engineering; environmental sciences including oceanography, atmospheric and earth sciences; mathematics and computer science including all fields of mathematics and computer-related sciences; and social and other behavioral sciences, including economics, political science, psychology, sociology. *Other Fields* are all Ph.D. degrees other than those awarded in the sciences and engineering.

SOURCE: National Science Foundation, Division of Policy Research and Analysis. Database: CASPAR. Some of the data within this database are estimates, incorporated where there are discontinuities within data series or gaps in data collection. Primary data source: U.S. Department of Education, National Center for Education Statistics, Higher Education General Information Survey (HEGIS): Degrees and Other Formal Awards Conferred.

S&E Ph.D. Degrees: Academic Field

During the past three decades, significant shifts have occurred in the proportion of Ph.D. degrees awarded among fields of study in science and engineering. During the 1960s, they increased in all broad fields, yet the field share shifted because of the relatively larger growth in engineering. In the 1970s, Ph.D. degrees in the social and behavioral, and life sciences continued to grow, while those in the physical sciences and engineering declined. In the 1980s, engineering Ph.D. production showed a relative resurgence.

Figure 2-94: Ph.D. Degrees Awarded in S&E by Field of Study

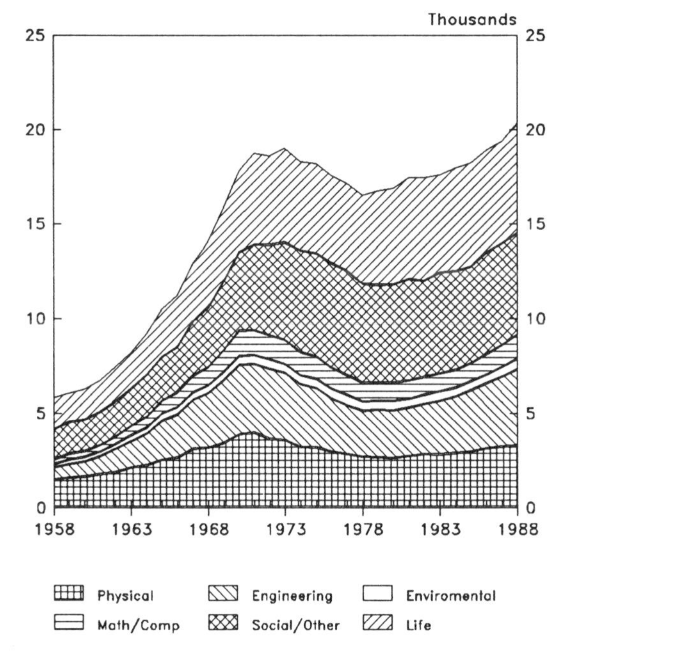

Figure 2-95: Distribution of Ph.D. Degrees Awarded in S&E by Field of Study

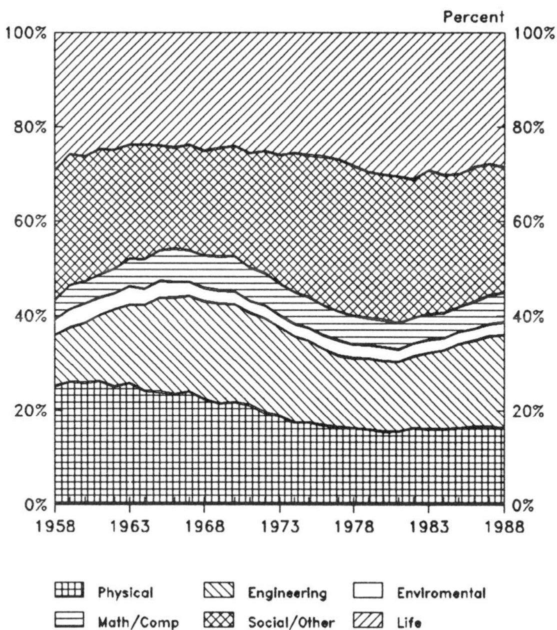

NOTE: Data series within the figures are not overlapped; top line represents total.

DEFINITION OF TERMS: *Physical* sciences include astronomy, chemistry, and physics. *Engineering* includes aeronautical and astronautical, chemical, civil, electrical, and mechanical engineering. *Environmental* sciences include oceanography, atmospheric, and earth sciences. *Mathematics/Computer* science include all fields of mathematics and computer-related sciences. *Social/Other* science include economics, political science, psychology, and sociology. *Life* sciences include agricultural, biological, medical, and other health sciences. S&E Ph.D.s include all those awarded from any academic institution.

SOURCE: National Science Foundation, Division of Policy Research and Analysis. Database: CASPAR. Some of the data within this database are estimates, incorporated where there are discontinuities within data series or gaps in data collection. Primary data source: U.S. Department of Education, National Center for Education Statistics, Higher Education General Information Survey (HEGIS): Degrees and Other Formal Awards Conferred.

S&E Ph.D. Degrees: Institution Governance

For public academic institutions, annual Ph.D. production in the sciences and engineering nearly quadrupled during the 1960s and early 1970s--from 3,300 in 1958 to 12,500 in 1973--then declined to 11,100 by 1978, rising to 13,600 by 1988. For private academic institutions, the annual Ph.D. production in the sciences and engineering nearly tripled during the 1960s and early 1970s, from 2,500 in 1958 to 6,500 in 1973, then declined to 5,300 by 1978, rising to 6,600 by 1988.

Figure 2-96: Science and Engineering Ph.D. Degrees by Institution Governance

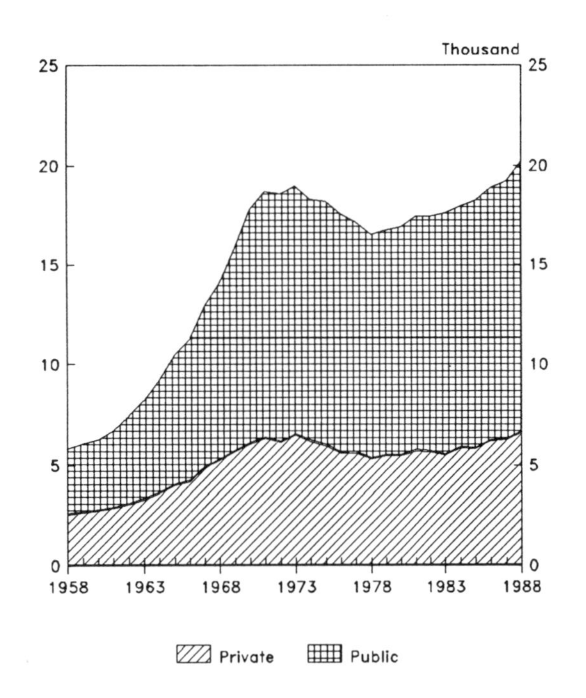

Figure 2-97: Distribution of Science and Engineering Ph.D. Degrees by Institution Governance

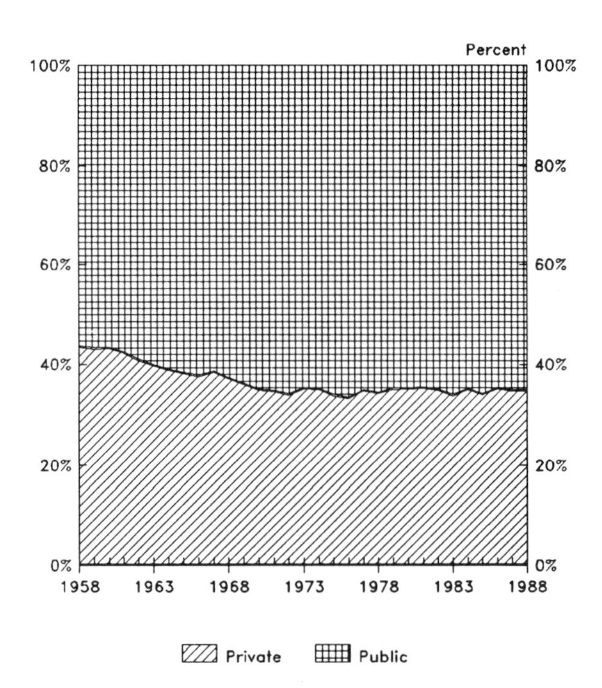

NOTE: Data series within the figures are not overlapping; top line represents total.

DEFINITION OF TERMS: Science and engineering Ph.D degrees include the following fields: Life sciences, including agricultural, biological, medical, and other health sciences; physical sciences including astronomy, chemistry, and physics; engineering including aeronautical and astronautical, chemical, civil, electrical, and mechanical engineering; environmental sciences including oceanography, atmospheric and earth sciences; mathematics and computer science including all fields of mathematics and computer-related sciences; and social and other sciences include economics, political science, psychology, sociology. Academic institutions offering Ph.D.s in the sciences and engineering include (1) all doctoral institutions, 116 public and 69 private, which have granted an average of 10 or more Ph.D. degrees per year in the natural sciences or engineering over the past two decades and (2) several of the 370 public and 854 private comprehensive institutions, which grant at least half of their degrees for courses of study that normally require 4 or more years to complete. *Public* institutions include higher education institutions under the control of--or affiliated with--federal, state, local, state and local, or state-related agencies. *Private* institutions are higher education institutions under the control of--or affiliated with--non-profit, independent organizations with no religious affiliation, or non-profit organizations with a religious affiliation.

SOURCE: National Science Foundation, Division of Policy Research and Analysis. Database: CASPAR. Some of the data within this database are estimates, incorporated where there are discontinuities within data series or gaps in data collection. Primary data source: U.S. Department of Education, National Center for Education Statistics, Higher Education General Information Survey (HEGIS): Degrees and Other Formal Awards Conferred.

S&E Ph.D. Degrees: Gender

The share of all science and engineering Ph.D. degrees awarded to women increased from 5 percent in 1958 to 30 percent by 1978, where it has remained during the 1980s. This increase results from a growing number of female Ph.D.s in the life, social and behavioral sciences during the 1970s and 1980s and a leveling off of Ph.D. degrees obtained by men during the same period.

Figure 2-98: Ph.D. Degrees Awarded in S&E by Gender

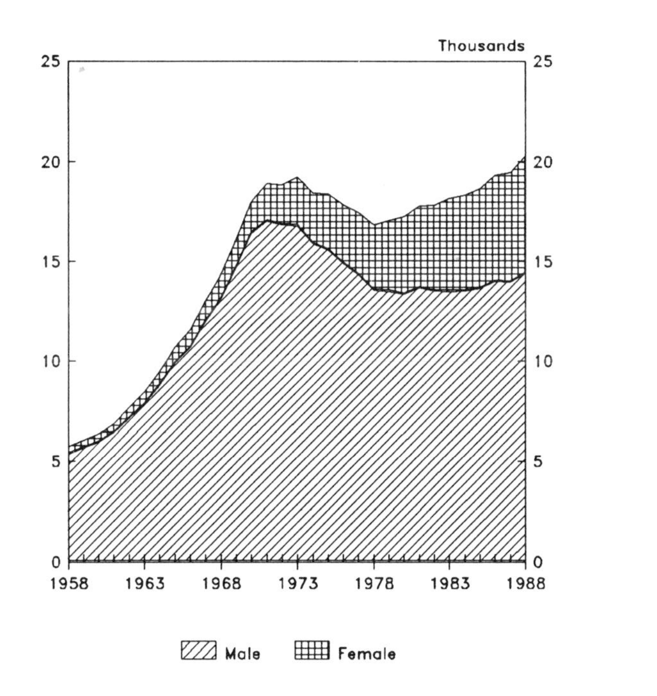

Figure 2-99: Distribution of Ph.D. Degrees Awarded in S&E by Gender

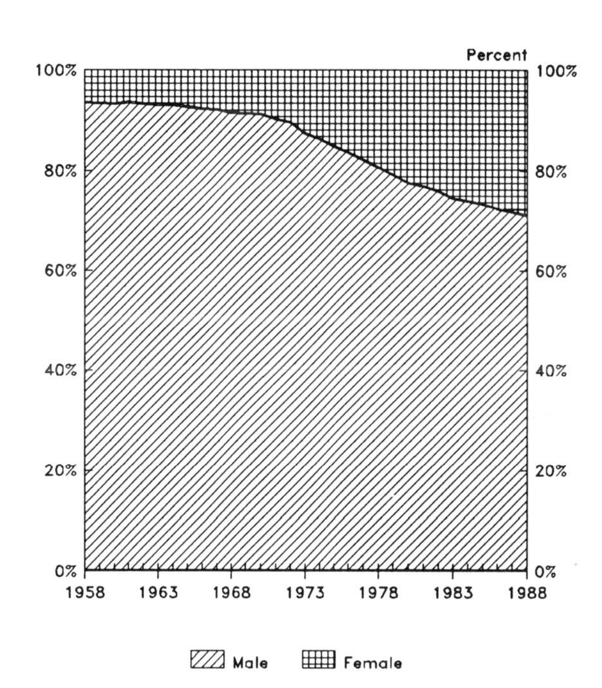

NOTE: Data series within the figures are not overlapped; top line represents total.

DEFINITION OF TERMS: Science and engineering Ph.D degrees are awarded in life sciences, including agricultural, biological, medical, and other health sciences; physical sciences including astronomy, chemistry, and physics; engineering includes aeronautical and astronautical, chemical, civil, electrical, and mechanical engineering; environmental sciences includes oceanography, atmospheric, and earth sciences; mathematics and computer science includes all fields of mathematics and computer-related sciences; and social and other behavioral sciences including economics, political science, psychology, sociology.

SOURCE: National Science Foundation, Division of Policy Research and Analysis. Database: CASPAR. Some of the data within this database are estimates, incorporated where there are discontinuities within data series or gaps in data collection. Primary data source: U.S. Department of Education, National Center for Education Statistics, Higher Education General Information Survey (HEGIS): Degrees and Other Formal Awards Conferred; National Science Foundation, Division for Science Resources Studies, Survey of Recent Science and Engineering Graduates.

Ph.D. Degrees by Ethnicity: Natural Sciences

During the past 15 years, the share of natural sciences Ph.D. degrees (U.S. citizens and permanent residents) obtained by minority students--Black, Hispanic, and Native American--has increased little, from 2 percent in 1973 to 4 percent by 1988. Asian-Americans have maintained a 6 percent share. The share obtained by white students declined from 93 percent in 1973 to 90 percent by 1988.

Figure 2-100: Ph.D. Degrees Awarded in Natural Sciences by Ethnicity

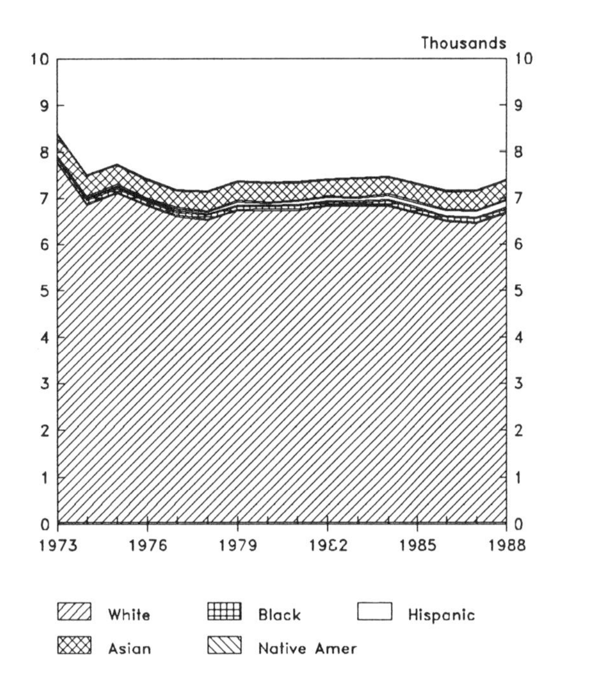

Figure 2-101: Distribution of Ph.D. Degrees Awarded in Natural Sciences by Ethnicity

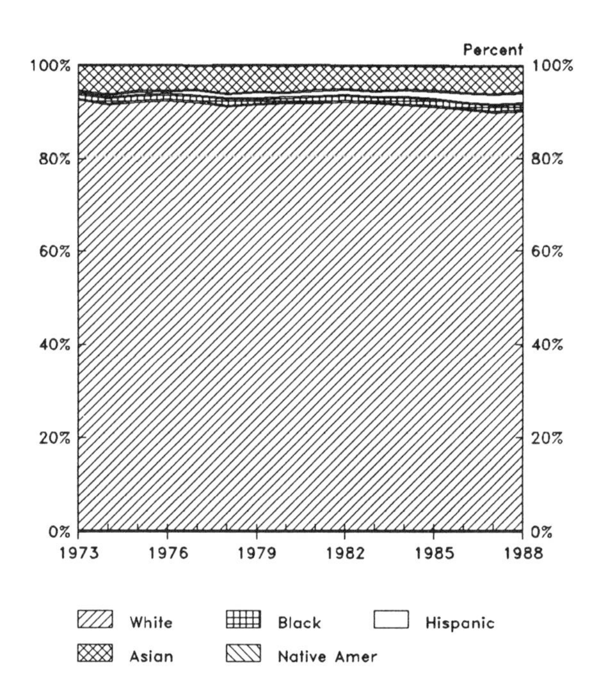

NOTE: Data series within the figures are not overlapping; top line represents total. Data include only U.S. citizens and permanent residents.

DEFINITION OF TERMS: Natural sciences Ph.D degrees include the following fields: Life sciences, including agricultural, biological, medical, and other health sciences; physical sciences including astronomy, chemistry, and physics; environmental sciences including oceanography, atmospheric and earth sciences; and mathematics and computer science including all fields of mathematics and computer-related sciences. U.S. citizens include all native-born or naturalized citizens of the United States. Permanent residents include all aliens residing within the United States on a permanent visa. *White* includes persons with origins in any of the orginal peoples of Europe, North Africa, or the Middle-East, except those of Hispanic origin; white also includes persons of unknown ethnicity. *Black* includes non-Hispanic persons with origins in any of the original black racial groups in Africa. *Hispanic* includes persons of Mexican, Cuban, Puerto Rican, Central or South American, or Spanish culture or origin, regardless of race. *Asian American* includes all persons with origins in any of original peoples of the Far East, Southeast Asia, the Indian subcontinent or Pacific Islands. *Native American* includes persons with origins in any of the orginal peoples of North America, including Alaskan Natives, maintaining cultural identification through tribal affiliation.

SOURCE: National Science Foundation, Division of Policy Research and Analysis. Database: CASPAR. Some of the data within this database are estimates, incorporated where there are discontinuities within data series or gaps in data collection. Primary data source: National Science Foundation, Division of Science Resource Studies, Survey of Recent Science and Engineering Graduates.

Ph.D. Degrees by Ethnicity: Engineering

During the past 15 years, the share of engineering Ph.D. degrees (U.S. citizens and permanent residents) obtained by minority students--Black, Hispanic, and Native American--increased from 1 percent in 1973 to 4 percent by 1988. The share of Asian-Americans increased from 12 percent in 1974 to 19 percent in 1979, averaged around 17 percent during the 1980s, with 16 percent in 1988. The share obtained by white students declined from 87 percent in 1973 to 80 percent by 1980, where it has remained.

Figure 2-102: Ph.D. Degrees Awarded in Engineering by Ethnicity

Figure 2-103: Distribution of Ph.D. Degrees Awarded in Engineering by Ethnicity

NOTE: Data series within the figures are not overlapping; top line represents total. Data include only U.S. citizens and permanent residents.

DEFINITION OF TERMS: Engineering Ph.D degrees include the following fields: aeronautical and astronautical, chemical, civil, electrical, and mechanical engineering. U.S. citizens include all native-born or naturalized citizens of the United States. Permanent residents include all aliens residing within the United States on a permanent visa. *White* includes persons with origins in any of the orginal peoples of Europe, North Africa, or the Middle-East, except those of Hispanic origin; white also includes persons of unknown ethnicity. *Black* includes non-Hispanic persons with origins in any of the original black racial groups in Africa. *Hispanic* includes persons of Mexican, Cuban, Puerto Rican, Central or South American, or Spanish culture or origin, regardless of race. *Asian American* includes all persons with origins in any of original peoples of the Far East, Southeast Asia, the Indian subcontinent or Pacific Islands. *Native American* includes persons with origins in any of the orginal peoples of North America, including Alaskan Natives, maintaining cultural identification through tribal affiliation.

SOURCE: National Science Foundation, Division of Policy Research and Analysis. Database: CASPAR. Some of the data within this database are estimates, incorporated where there are discontinuities within data series or gaps in data collection. Primary data source: National Science Foundation, Division of Science Resource Studies, Survey of Recent Science and Engineering Graduates.

Ph.D. Degrees by Citizenship: Natural Sciences

Since 1978, the share of Ph.D. degrees in natural sciences awarded to foreign students with temporary U.S. visas grew from 14 percent to 24 percent by 1988.

Figure 2-104: Ph.D. Degrees Awarded in Natural Sciences by Citizenship

Thousands

US Citizens
Non-US (Perm Visa)
Non-US (Temp Visa)

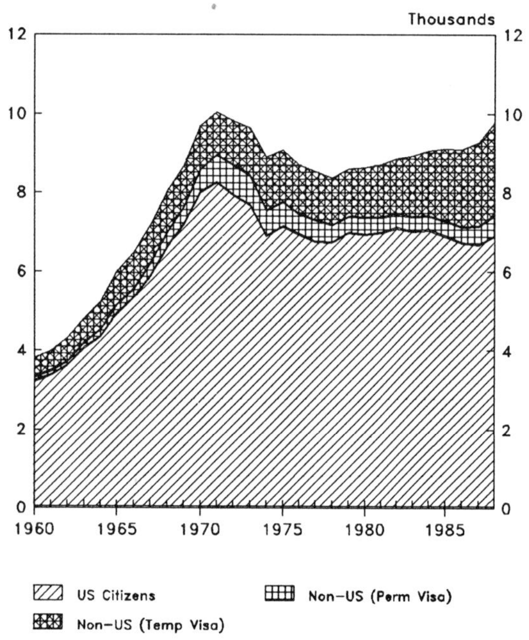

Figure 2-105: Distribution of Ph.D. Degrees Awarded in Natural Sciences by Citizenship

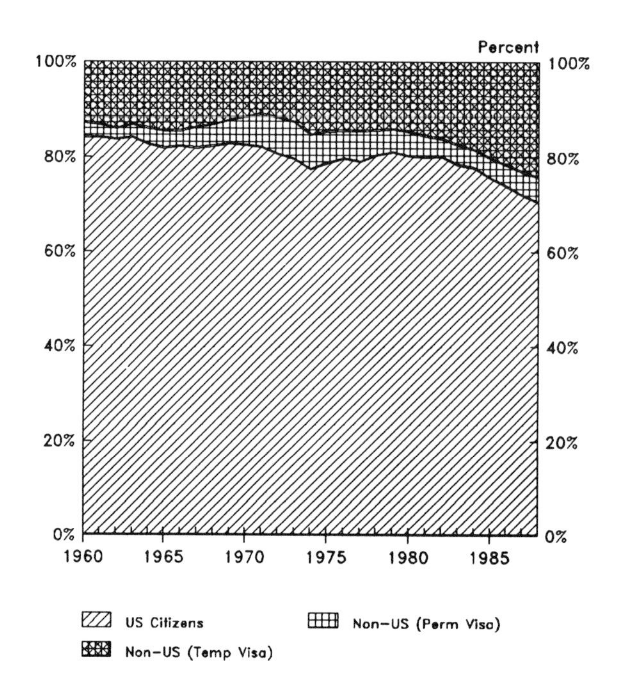

NOTE: Data series within the figures are not overlapped; top line represents total.

DEFINITION OF TERMS: Natural Sciences Ph.D degrees include the following fields: life sciences, biological, medical, and other health sciences; physical sciences including astronomy, chemistry, and physics; environmental sciences including oceanography, atmospheric, and earth sciences; and mathematics and computer science includes all fields of mathematics and computer-related sciences. *U.S. Citizens* include native-born or naturalized citizens of the United States. *Non-U.S. (Permanent Visa)* includes all aliens residing in the United States on a permanent visa. *Non-U.S.(Temporary Visa)* includes all aliens residing in the United States on a temporary visa.

SOURCE: National Science Foundation, Division of Policy Research and Analysis. Database: CASPAR. Some of the data within this database are estimates, incorporated where there are discontinuities within data series or gaps in data collection. Primary data source: U.S. Department of Education, National Center for Education Statistics, Higher Education General Information Survey (HEGIS): Degrees and Other Formal Awards Conferred; National Science Foundation, Division for Science Resources Studies, Survey of Recent Science and Engineering Graduates.

Ph.D. Degrees by Citizenship: Engineering

The share of engineering Ph.D. degrees awarded to foreign students with temporary U.S. visas grew from 16 percent in 1958 to 30 percent by 1978, then rapidly increased to nearly 45 percent by 1985.

Figure 2-106: Ph.D. Degrees Awarded in Engineering by Citizenship

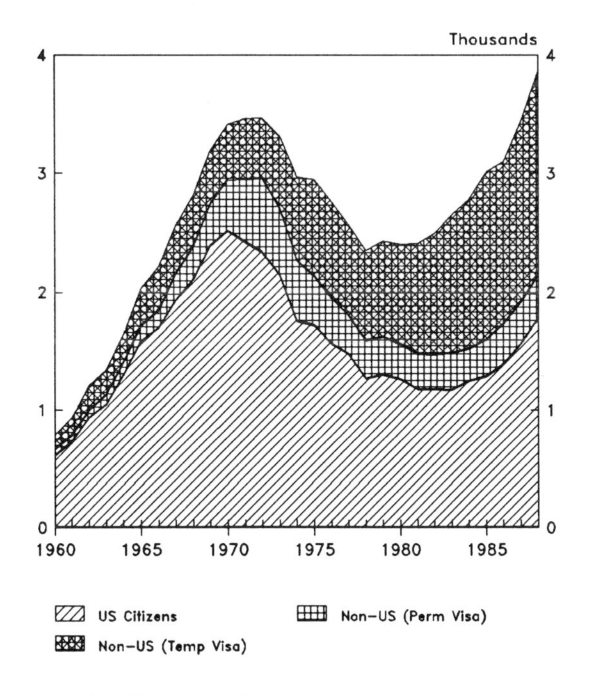

Figure 2-107: Distribution of Ph.D. Degrees Awarded in Engineering by Citizenship

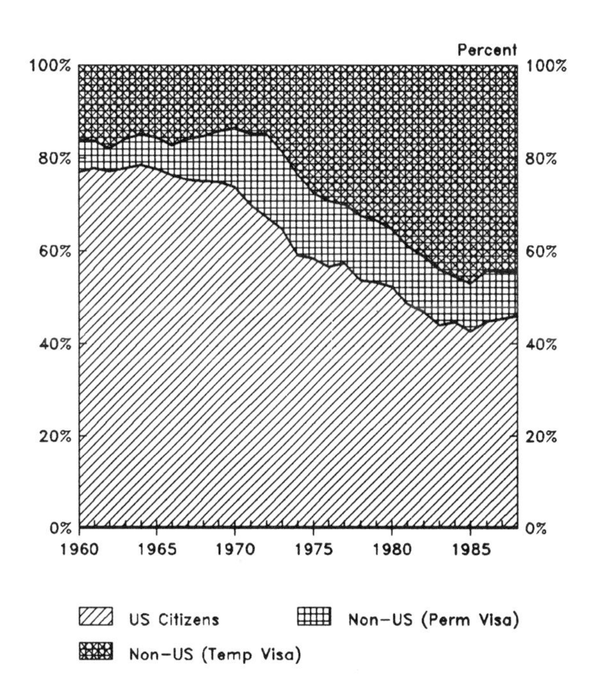

DEFINITION OF TERMS: Engineering Ph.D degrees include the following fields: aeronautical and astronautical, chemical, civil, electrical, and mechanical engineering. *U.S. Citizens* include all native or naturalized citizens of the United States. *Non-U.S.-Permanent Visa* includes all aliens residing within the United States with a permanent visa. *Non-U.S. Temporary Visa* includes all aliens residing within the United States with a temporary visa.

SOURCE: National Science Foundation, Division of Policy Research and Analysis. Database: CASPAR. Some of the data within this database are estimates, incorporated where there are discontinuities within data series or gaps in data collection. Primary data source: U.S. Department of Education, National Center for Education Statistics, Higher Education General Information Survey (HEGIS): Degrees and Other Formal Awards Conferred.

PRIMARY DATA SOURCES

National Research Council, Office of Scientific and Engineering Personnel, Survey of Doctoral Recipients.

National Science Foundation, Division of Science Resource Studies, National Survey of Natural and Social Scientists and Engineers.

National Science Foundation, Division of Science Resource Studies, Survey of Earned Doctorates Awarded in the United States.

National Science Foundation, Division of Science Resource Studies, Survey of Scientific and Engineering Personnel Employed at Universities and Colleges.

National Science Foundation, Division of Science Resource Studies, Survey of Federal Funds for Research and Development.

National Science Foundation, Division of Science Resource Studies, Survey of Federal Support to Universities, Colleges, and Non-profit Organizations.

National Science Foundation, Division of Science Resource Studies, Survey of Scientific and Engineering Expenditures at Universities and Colleges.

National Science Foundation, Division of Science Resource Studies, Survey of Industrial Research and Development.

U.S. Department of Commerce, Bureau of Economic Analysis, Survey of Current Business and Commerce.

U.S. Department of Commerce, Bureau of the Census, Current Population Reports, Projections of the Population of the United States by Age, Sex, and Race: 1983-2080.

U.S. Department of Education, National Center for Education Statistics, Higher Education General Information Survey (HEGIS): Degrees and Other Formal Awards Conferred.

U.S. Department of Education, National Center for Education Statistics, Higher Education General Information Survey (HEGIS): Fall Enrollment in Institutions of Higher Education.

U.S. Department of Education, National Center for Education Statistics, Higher Education General Information Survey (HEGIS): Salaries, Tenure, and Fringe Benefits of Full-time Instructional Faculty.

U.S. Department of Education, National Center for Education Statistics, Higher Education General Information Survey (HEGIS): Financial Statistics of Institutions of Higher Education.